Maritime Pipefitting
Level Two

Trainee Guide

PEARSON

Boston   Columbus   Indianapolis   New York   San Francisco   Upper Saddle River
Amsterdam   Cape Town   Dubai   London   Madrid   Milan   Munich   Paris   Montreal   Toronto
Delhi   Mexico City   São Paulo   Sydney   Hong Kong   Seoul   Singapore   Taipei   Tokyo

NCCER
President: Don Whyte
Director of Product Development: Daniele Dixon
Maritime Pipefitting Project Manager: Patty Bird
Senior Manager: Tim Davis
Quality Assurance Coordinator: Debie Ness

Desktop Publishing Coordinator: James McKay
Permissions Specialist: Amanda Werts
Production Specialist: Megan Casey
Editor: Chris Wilson

Writing and development services provided by Topaz Publications, Liverpool, NY
Lead Writer/Project Manager: Tom Burke
Desktop Publisher: Joanne Hart
Art Director: Alison Richmond

Permissions Editors: Toni Burke, Andrea LaBarge
Writers: Tom Burke, Troy Staton, Veronica Westfall

Pearson Education, Inc.
Editorial Director: Vernon R. Anthony
Executive Editor: Alli Gentile
Editorial Assistant: Douglas Greive
Program Manager: Alexandrina B. Wolf
Operations Supervisor: Deidra M. Skahill
Art Director: Jayne Conte
Director of Marketing: David Gesell
Executive Marketing Manager: Derril Trakalo
Marketing Manager: Brian Hoehl
Marketing Coordinator: Crystal Gonzalez

Composition: NCCER
Printer/Binder: Document Technology Resources, Fredericksburg, VA
Cover Printer: Document Technology Resources, Fredericksburg, VA
Text Fonts: Palatino and Univers

Credits and acknowledgments for content borrowed from other sources and reproduced, with permission, in this textbook appear at the end of each module.

10 9 8 7 6 5 4 3 2 1

PEARSON

Perfect bound   ISBN-13: 978-0-13-340478-4
                 ISBN-10:   0-13-340478-1

Preface

To the Trainee

Congratulations! If you're training under an NCCER-Accredited Training Sponsor, you have likely completed *Maritime Pipefitting Level One*. Now you are well on your way to more specific skills in training in maritime pipefitting.

In *Maritime Pipefitting Level Two*, you will learn about piping systems; pipe fabrications; brazing; identifying valves, flanges, and gaskets; and drawing and detail sheets. These areas will provide you with knowledge and skills to advance your career.

The maritime industry is facing an aging workforce and shortage of skilled workers. This shortage extends to the pipefitting trade. Skilled pipefitters are in high demand in almost every shipyard. By completing this training you are gaining the abilities needed to help fill that demand. We hope you continue your training beyond this textbook to further your knowledge and skills in this field.

If you're training through an NCCER-Accredited Training Program Sponsor and successfully pass the module exams and performance tests in this course, you may be eligible for credentialing through NCCER's National Registry. Check with your instructor or local program sponsor to find out. To learn more, go to **www.nccer.org** or contact us at 1.888.622.3720.

We invite you to visit the NCCER website at **www.nccer.org** for information on the latest product releases and training, as well as online versions of the *Cornerstone* newsletter and Pearson's NCCER product catalog.

Your feedback is welcome. You may email your comments to **curriculum@nccer.org** or send general comments and inquiries to **info@nccer.org**.

NCCER Standardized Curricula

NCCER is a not-for-profit 501(c)(3) education foundation established in 1996 by the world's largest and most progressive construction companies and national construction associations. It was founded to address the severe workforce shortage facing the industry and to develop a standardized training process and curricula. Today, NCCER is supported by hundreds of leading construction and maintenance companies, manufacturers, and national associations. The NCCER Standardized Curricula was developed by NCCER in partnership with Pearson, the world's largest educational publisher.

Some features of the NCCER Standardized Curricula are as follows:

- An industry-proven record of success
- Curricula developed by the industry for the industry
- National standardization providing portability of learned job skills and educational credits
- Compliance with the Office of Apprenticeship requirements for related classroom training (*CFR 29:29*)
- Well-illustrated, up-to-date, and practical information

NCCER also maintains a National Registry that provides transcripts, certificates, and wallet cards to individuals who have successfully completed modules of NCCER's Curricula. *Training programs must be delivered by an NCCER Accredited Training Sponsor in order to receive these credentials.*

Objectives

When you have completed this module, you will be able to do the following:

1. Identify and explain the types of piping systems used in maritime applications and the methods used to identify them.
2. Describe the types of pipe used in maritime systems, how they are used, and how they are sized.
3. Explain the effects and corrective measures for thermal expansion and heat loss in piping systems.

Performance Tasks

Under the supervision of the instructor, you should be able to do the following:

1. Identify the type of piping system designated by color codes.
2. Identify piping systems by material.
3. Identify pipe schedules by pipe.

Trade Terms

Acid
Butt weld
Caustic
Condensate
Cupronickel

Material safety data sheet (MSDS)
Pounds per square inch gauge (psig)
Socket weld
Water hammer

Industry-Recognized Credentials

If you are training through an NCCER-accredited sponsor, you may be eligible for credentials from NCCER's Registry. The ID number for this module is 85201-13. Note that this module may have been used in other NCCER curricula and may apply to other level completions. Contact NCCER's Registry at 888.622.3720 or go to **www.nccer.org** for more information.

Contents

Topics to be presented in this module include:

1.0.0 Introduction 1

2.0.0 Maritime Piping Systems 1

  2.1.0 Color Coding Used on Piping Systems 1

3.0.0 Types of Pipe 2

  3.1.0 Carbon Steel Pipe 2

    3.1.1 Schedules and Wall Thicknesses 2

  3.2.0 Copper Pipe 3

    3.2.1 Types and Sizes of Copper Pipe 3

    3.2.2 Copper Pipe Labeling 5

  3.3.0 Cupronickel and Nickel-Copper Pipe 5

  3.4.0 Stainless Steel Pipe 6

4.0.0 Working with Piping Systems 6

  4.1.0 Chemical Systems 6

  4.2.0 Compressed Air Piping Systems 6

  4.3.0 Fuel Oil and Flammable Liquid Piping Systems 7

  4.4.0 Steam Piping Systems 8

  4.5.0 Water Piping Systems 8

5.0.0 Thermal Expansion of Pipes 9

  5.1.0 Flexibility in Layout 10

  5.2.0 Installing Expansion Loops or Expansion Joints 10

  5.3.0 Cold Springing 11

6.0.0 Pipe Insulation 12

Figures and Tables

Figure 1   Pipe used in maritime applications...................................1
Figure 2   Fire suppression piping 2
Figure 3   Butt weld pipe joint 2
Figure 4   Inside and outside diameters of pipe 3
Figure 5   Annealed (soft) and hard-drawn copper pipe 4
Figure 6   Comparison of copper pipe types...................................5
Figure 7   Identifying markings on Type ACR pipe 5
Figure 8   Compressed air piping system......................................7
Figure 9   Fuel oil piping system 8
Figure 10   Steam piping system 9
Figure 11   Potable water piping system 9
Figure 12   Cooling water piping system......................................10
Figure 13   Expansion joint and expansion loops..............................10
Figure 14   Flexibility in layout 11
Figure 15   Expansion loop 11
Figure 16   Cold springing 12
Figure 17   Insulated pipes 12

Table 1   Color Codes Used to Identify Piping...............................1
Table 2   Carbon Steel Pipe Dimensions......................................4
Table 3   Copper Pipe Color Codes...5
Table 4   Expansion of Different Types of Pipe 10
Table 5   Insulating Materials for Different Temperature Applications 13

1.0.0 INTRODUCTION

A piping system is a complete network of pipes, fittings, valves, and other components that are designed to work together to convey specific material. Piping systems must be designed and fabricated according to codes, plans, specifications, and component installation instructions. Where conflicts between these sources are found, always use the method that is the most conservative and provides the greatest margin of safety. This module explains the different types of piping systems used on ship and offshore rigs, along with thermal expansion of pipes, and the use of pipe insulation. There are many different types and sizes of pipe (*Figure 1*). Pipefitters must become familiar with the different types, their uses, and the joining methods used with them.

Keep in mind that pipefitters are not permitted to make judgment calls regarding the type of pipe to use in a given situation. Just because polyvinyl chloride (PVC) is used for potable water in one location, for example, does not mean it can be used for that purpose in another. Potable water is water that is safe to drink. The project specification always governs. If there is any doubt, ask a supervisor.

2.0.0 MARITIME PIPING SYSTEMS

The sheer number of different products used on ships and offshore rigs dictates that the vessel will have many piping systems. Such systems include the following:

- Freshwater for cooking, bathing, drinking, and cooling
- Wastewater system to handle sewage and other waste

- Seawater handling for bilges, ballast systems, and engine cooling
- Fire suppression systems, including water, CO_2, foam, dry powder, and inert gas
- Petroleum products, including diesel fuel, lubricants, and hydraulic fluid on ships and crude oil on offshore rigs
- Pneumatic and compressed air systems
- Low-pressure and high-pressure systems.
- Steam piping for heating systems
- Hydraulic systems, including steering gear, watertight doors, and ramp control systems
- Piping systems to handle liquid and gaseous cargo, such as that on tankers
- Offshore rigs require additional piping systems to handle crude and refined oil, as well as systems for handling drilling mud, cement, and gray water

2.1.0 Color Coding Used on Piping Systems

Imagine walking into a space that has 100 pipes running through. How do you find the one you are looking for? Part of the answer to that question lies in the use of color codes for piping. Standards have been developed for painting or otherwise marking pipe to represent the type of material it carries and the piping system it belongs to. *Table 1* lists the common color codes. Red is the universal color for fire suppression piping and fixtures, for example (see *Figure 2*). Any location may have a number of pipes with the same color code. Therefore, in addition to color codes, each piping run is marked with an identifier that links it to a drawing so that it can be precisely located.

Table 1 Color Codes Used to Identify Piping

Color	Piping System Examples
Orange	Lubricating oil, hydraulic oil
Red	Fire suppression
Black	Wastewater, bilge water
Brown	Fuel – diesel, aviation, heavy fuel
Yellow	Flammable gases – hydrogen, acetylene, liquid propane
Green	Seawater – ballast water, cooling water
Silver	Steam
White	Air – ventilation, exhaust, supply air
Blue	Water – freshwater, chilled water, potable water
Grey	Air – oxygen, inert gases, refrigerant, breathing air
Violet	Acids

85201-13_F01.EPS

Figure 1 Pipe used in maritime applications.

Figure 2 Fire suppression piping.

3.0.0 TYPES OF PIPE

Galvanized mild steel or carbon steel pipe are common types of piping used on ships and offshore rigs. However, piping material varies by client. The US Navy for example, uses very little steel pipe on their ships. Mild steel pipe can be joined with threaded fittings, grooved fittings, or by welding using either the socket weld or butt weld (*Figure 3*) method. Polyvinyl chloride (PVC) and chlorinated polyvinyl chloride (CPVC) plastic pipe, copper, fiberglass, stainless steel, and cupronickel pipe are also used in certain applications. For example, Schedule 80 plastic pipe may be used for black water (sewage waste) or potable (fresh, drinkable) water. Fiberglass pipe is becoming more common in systems handling seawater and other corrosive substances. Carbon steel, copper, cupronickel, and stainless steel pipe are described in this section. Plastic and fiberglass pipe are covered in a separate module.

Pipes that carry dangerous chemicals or particularly corrosive fluids are commonly made from stainless steel. Some chemicals can be carried only in stainless steel cargo tanks and pipes. Copper pipe is used in some seawater applications, but cannot be used in systems where the liquid

Figure 3 Butt weld pipe joint.

temperature exceeds 200°F (93°C). Fiberglass is often used for ballast and brine pipes as well as seawater cooling systems. It is used in sizes ranging from 2 to 14 inches. Vacuum-type sewage systems may be made from plastic with plastic welded joints. PVC and CPVC may also be used in potable water and water-making systems. CPVC is able to handle higher pressure than PVC. The use of plastic pipe for other applications is limited because the pipe must pass a fire-resistance test. Plastic pipe will not pass the most restrictive level of fire resistance, so it cannot be used in some locations.

Corrosion is the main source of concern in all types of piping. This is especially true in systems that handle seawater, which has about three times the corrosive action of freshwater. Piping systems need to be inspected for corrosion because excessive corrosion will eventually result in pipe failure. Corrosion is most likely to occur around joints and bends.

3.1.0 Carbon Steel Pipe

Carbon steel pipe is widely used in maritime piping systems because it is ductile, durable, machinable, and less expensive than most other types of pipe. Carbon steel is more susceptible to corrosion than stainless steel, fiberglass, or cupronickel. Almost all carbon steel pipe is galvanized to protect it from corrosion. Galvanized pipe is made by dipping carbon steel pipe into a bath of molten zinc. The zinc coats the inside and outside of the pipe. Galvanized pipe is used to carry water, air, and other fluids.

Carbon steel pipe is listed by its nominal size in inches. For sizes of pipe up to and including 12", nominal size is an approximation of the inside diameter of Schedule 40 pipe, which is the standard weight pipe. From 14" on, nominal size reflects the outside diameter of the pipe. *Figure 4* shows how the inside and outside diameters of a pipe are determined. Note that the outside diameter remains the same, while the inside diameter changes as the pipe schedule changes.

3.1.1 Schedules and Wall Thicknesses

Wall thickness can be described in two ways. The first is by schedule. As the schedule numbers increase, the wall thickness gets larger; therefore, the pipe is stronger and can withstand more pressure. Schedule numbers for pipe range from 5 to 160, but no pipe smaller than Schedule 40 should be threaded. It is important to remember that the schedule number refers only to the wall thickness

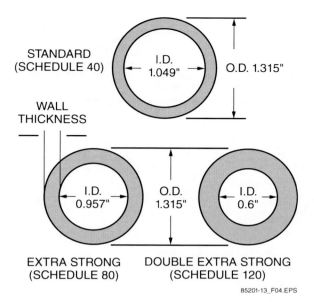

**STANDARD
(SCHEDULE 40)**

I.D. 1.049"   O.D. 1.315"

**WALL
THICKNESS**

I.D. 0.957"   O.D. 1.315"   I.D. 0.6"

**EXTRA STRONG
(SCHEDULE 80)**   **DOUBLE EXTRA STRONG
(SCHEDULE 120)**

85201-13_F04.EPS

Figure 4 Inside and outside diameters of pipe.

of a pipe of a given nominal size. A ¾" Schedule 40 pipe will not have the same wall thickness as a 1" Schedule 40 pipe. Another way to describe pipe wall thickness is by manufacturer's weight. From smallest to largest, there are three classifications in common use today:

- *STD* – Standard wall
- *XS* – Extra-strong wall
- *XXS* – Double extra-strong wall

Schedule numbers and wall thicknesses are somewhat interchangeable. Schedule 40 galvanized pipe and standard-wall galvanized pipe have the same wall thickness for all sizes up to and including 10" nominal size. Schedule 80 and extra-strong wall galvanized pipe have the same wall thickness through 8" nominal size. Common wall thicknesses of carbon steel pipe are Schedule 40, 80, and 160, and double extra-strong. Common wall thicknesses of stainless steel pipe range from Schedule 5 to 160. Common pipe dimensions are shown in *Table 2*.

3.2.0 Copper Pipe

There are two primary categories of copper pipe based on the temper, or hardness, of the material. Annealed copper pipe is often referred to as soft copper. It is generally sold in manageable rolls of various lengths. Both 25' and 50' rolls are common. Annealed pipe is very easy to bend and form. As long as it is done with the proper tools and technique, it can be formed without suffering significant damage. Hard-drawn pipe, commonly known as hard copper, is fabricated by drawing, or pulling, the copper through a series of dies to decrease its size to the desired diam-

eter. This process tends to harden the pipe, making it more rigid. Both annealed and hard-drawn pipe are shown together in *Figure 5*. As a general rule, the working pressure rating of hard copper is about 60 percent higher than that of soft copper. Hard copper pipe is sold in straight lengths, usually 20 feet in length. Brazing is the most common method for joining both soft and hard copper.

3.2.1 Types and Sizes of Copper Pipe

Different types of copper pipe are identified with one or more letters. The primary difference in each type is the wall thickness. A copper pipe size of 1" will be used in this section to demonstrate the variance in wall thickness (see *Figure 6*).

The sizing of most copper pipe is based on the nominal (standard) size of the inside diameter. This is true for Types K, L, M, and DWV. However, remember that each of these types have different wall thicknesses, and the inside diameter (ID) of the pipe changes as the wall thickness changes. The OD remains the same. This means that the OD of Type M, Type L, and Type K pipe is the same for any given size. The ID of each one, however, is different.

Type M copper pipe has the thinnest wall of the commonly used products. It is also the lightest in weight and lowest in cost as a result. Type M is typically used in domestic plumbing systems for water, and is suitable for many other low pressure and drainage applications. A 1" Type M copper pipe has a wall thickness of 0.035".

Type L copper pipe has a thicker wall and can be used in higher-pressure applications than Type M. A 1" Type L copper pipe has a wall thickness of 0.050".

Type K pipe has the thickest wall and, of course, is the heaviest in weight per foot. Type K is typically used in the most challenging and critical applications. Type DWV has thin walls. The acronym DWV stands for drain, waste, and vent. It is not designed for pressurized applications. Since it is primarily designed for DWV service, the smallest size available is 1¼". This size has a wall thickness of 0.040". Due to the cost of the material compared to plastic options, it is typically only used in DWV applications where an attractive appearance is desired. There is little incentive to use it where the installation is hidden from public view.

Type ACR pipe is the type used for refrigerant lines in heating, ventilation, and air conditioning (HVAC) systems. It is quite different from other types of copper pipe. ACR is an acronym for air conditioning and refrigeration. This pipe was specifically developed for refrigerant piping

Table 2 Carbon Steel Pipe Dimensions

Nominal Size	Outside Diameter	Inside Diameter			
		Standard, Schedule 40	Extra-Strong, Schedule 80	Schedule 160	Double Extra-Strong
⅛	0.405	0.269	0.215	–	–
¼	0.540	0.364	0.302	–	–
⅜	0.675	0.493	0.423	–	–
½	0.840	0.622	0.546	0.466	0.252
¾	1.050	0.824	0.742	0.614	0.434
1	1.315	1.049	0.957	0.815	0.599
1¼	1.660	1.380	1.278	1.160	0.896
1½	1.900	1.610	1.500	1.338	1.100
2	2.375	2.067	1.939	1.689	1.503
2½	2.875	2.469	2.323	1.885	1.771
3	3.500	3.068	2.900	2.625	2.300
3½	4.000	3.548	3.364	–	–
4	4.500	4.026	3.826	3.438	3.152
5	5.562	5.047	4.813	4.313	4.063
6	6.625	6.065	5.761	5.187	4.897
8	8.625	7.981	7.625	6.813	6.875

85201-13_T02.EPS

85201-13_F05.EPS

Figure 5 Annealed (soft) and hard-drawn copper pipe.

use. Type ACR is available in both hard and soft versions. This type is the same in construction as Type L, with the same dimensions. What sets it apart is the fact that it is purged of all air, and then charged with nitrogen after it is manufactured. It is then tightly capped to maintain the nitrogen atmosphere inside at a very low pressure. This eliminates the presence of oxygen inside the pipe, thereby eliminating copper oxides. When a new length of Type ACR is first opened, it will be extremely clean, bright, and shiny inside. This

level of cleanliness is very important to the refrigeration circuit. The caps or plugs should always be replaced on the unused portion of ACR pipe once it is cut. Although any nitrogen is now gone and air has entered, replacing the plugs prevents the constant entry of fresh oxygen and debris.

One additional feature of importance regarding Type ACR is the way it is sized. While a length of Type L pipe would be called out as 1" (its nominal diameter), Type ACR of the same identical size would be called 1⅛" pipe (the actual outside diameter). Trade workers must be specific about the copper pipe type and size they need.

There are also medical gas types of copper pipe. They are typically identified as Medical Gas Type K and Medical Gas Type L.

Copper pipe is made to metric standards as well as imperial standards. *American Society for Testing and Materials (ASTM) Standard B88* covers imperial copper pipe standards. *ASTM Standard B88M* is a companion publication covering the metric sizes. Metric copper pipe has several different types as well, identified as Types A, B, and C. Type A has the thickest wall, while Type C has the thinnest wall. There are quite a few other standards to which copper pipe is manufactured across the world.

It is important to remember that the metric and imperial pipe sizes and fittings are not interchangeable. Unlike metric hardware, such as nuts

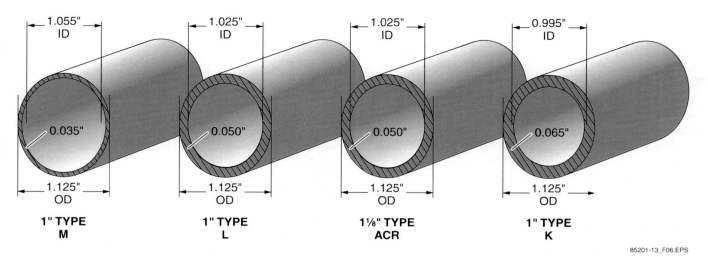

1" TYPE M — 1.055" ID, 0.035", 1.125" OD

1" TYPE L — 1.025" ID, 0.050", 1.125" OD

1⅛" TYPE ACR — 1.025" ID, 0.050", 1.125" OD

1" TYPE K — 0.995" ID, 0.065", 1.125" OD

85201-13_F06.EPS

Figure 6 Comparison of copper pipe types.

and bolts, metric copper pipe is virtually non-existent in North American products or installations. When it is encountered, it would be handled in the same manner as the imperial sizes. Remember that metric fittings would need to be matched with the metric pipe.

3.2.2 Copper Pipe Labeling

Manufacturers must permanently etch or stamp Types K, L, M, DWV, and medical gas copper pipe to show the pipe type, the name or trademark of the manufacturer, and the country of origin. This same information must be printed on hard-drawn pipes in a specific color that corresponds to the pipe type. Type ACR does not require these same permanent markings, but they may be present. Hard ACR pipe is best identified by the printed markings (*Figure 7*). Soft ACR pipe may not have printed or permanent markings at all. *Table 3* shows five types of copper pipe and the corre-

sponding color code for labeling. Note that the Type ACR color code is the same as Type L.

3.3.0 Cupronickel and Nickel-Copper Pipe

Cupronickel pipe, otherwise known as copper-nickel pipe, is made from a combination of copper (Cu) and nickel (Ni), with a small percentage of manganese and iron added for strength. The abbreviation CuNi is often used to designate this type of pipe. Copper is the majority metal. The

Table 3 Copper Pipe Color Codes

Type	Color Code
K	Green
L	Blue
M	Red
DWV	Yellow
ACR	Blue

85201-13_F07.EPS

Figure 7 Identifying markings on Type ACR pipe.

ratio of copper to nickel ranges from a high of 90-10 to 70-30. The 70-30 grade is commonly used in marine work. Cupronickel pipe is used in seawater applications because it resists corrosion caused by seawater. Cupronickel can be welded with methods that use a consumable welding electrode. Cupronickel pipe is available in the same sizes and schedules as carbon steel pipe.

A nickel-copper alloy known as Monel™ is also used in maritime work. This alloy is abbreviated NiCu. A 70-30 NiCu alloy is used in piping systems, pump shafts, and seawater valves.

3.4.0 Stainless Steel Pipe

Stainless steel pipe is made by alloying iron with a certain percentage of chromium and nickel. The alloys increase the resistance to corrosion, strengthen the steel, and make the steel able to withstand extreme temperatures. It is good for applications that require the maximum resistance to corrosion, and is therefore commonly used in seawater systems. It is also used for instrument lines and heat-transfer equipment. It is common on tankers that carry corrosive chemicals.

Stainless steel can be kept sterile to prevent contamination of food and dairy products so it is common in food service applications. Stainless steel is very expensive, so it is only used in critical applications that expose the pipe to high temperature, extreme low temperature, or high levels of corrosion. Stainless steel pipe must be inspected for pitting, especially when used in seawater applications. If pitting has occurred, rust will be visible on the outside of the pipe. It can be removed as a temporary fix by cleaning the affected area with a stainless steel wire brush and then painting over it.

4.0.0 WORKING WITH PIPING SYSTEMS

There are certain precautions that need to be taken by anyone working with and around piping systems. Some systems, such as hydraulic and pneumatic systems, are under pressure that could cause injury if released. Other systems contain flammable material such as diesel fuel. Piping in ships that transport liquid cargos may contain residual material that might be **caustic** or flammable. The following are some guidelines for working with these systems.

4.1.0 Chemical Systems

Acids, caustics, and other chemical substances can attack the piping, fittings, and valves through which they pass. Chemical piping systems can be either vapor systems or liquid systems. Follow these safety guidelines when working around chemical piping systems:

- Always know exactly what chemicals are contained in the piping systems and know the location of the **material safety data sheet (MSDS)** or safety data sheet (SDS) for all chemicals in the work area.

> **NOTE**
>
> The Globally Harmonized System of Classification and Labeling of Chemicals, known as GHS introduced a new safety data sheet (SDS) format containing 16 sections that will gradually replace the MSDS. An MSDS or SDS must be available for materials on the site that contain hazardous chemicals. The MSDS/SDS is available from the chemical supplier and from the Internet. All workers should read the MSDS/SDS for the materials they work with, both to protect the environment, and for their personal safety.

- Wear personal protective equipment as specified by the employer and the MSDS/SDS, including a chemical resistance suit.
- Always know the location of the nearest eyewash fountain and quick-drench shower.
- If someone inhales chemicals, move the person from the exposure area into fresh air, keep the person warm and at rest, and seek medical attention immediately.
- If a person is unconscious in a hazardous area, do not enter the hazardous area to retrieve that person. Contact the emergency response team immediately.

> **WARNING!**
>
> If you notice a gas or vapor cloud, notify operations immediately.

- Never walk through a vapor cloud in a hazardous area. Go upwind or walk around the cloud.

4.2.0 Compressed Air Piping Systems

Compressed air piping systems are classified as either instrument air or utility air. Air that operates many instruments is instrument air. Instrument air is compressed and dried to remove the moisture, because moisture in the air fouls or damages the equipment. The air is routed to all of the pneumatic instruments through piping systems. Instrument air piping systems are usually galvanized steel, stainless steel, or copper.

Utility air is air that is compressed and then used to drive pneumatic motors in power tools, to propel fluids, to operate various cleaning services, and to empty and dry piping. Most utility air systems are welded carbon steel pipe. Air pressure for each system varies, depending on the application, but usually ranges between 50 and 125 **pounds per square inch gauge (psig)**. There are uses for higher-pressure air, such as engine starting systems. But systems at 125 psi and lower are the most common. *Figure 8* shows a schematic example of a compressed air piping system.

Follow these safety guidelines when working around compressed air systems:

- Because compressed air is under pressure, precautions must be taken in the event of fire since excessive heat may cause vessels to rupture explosively.
- Never direct the airflow from an air nozzle directly onto skin because an air bubble could enter the skin and the bloodstream.
- Never stop a leak in a compressed air piping system with your finger or any other part of your body.

- Do not inhale compressed air because this may cause decompression sickness, which has a variety of adverse effects and can lead to long-term illness and even death.
- In case of inhalation, move the person to fresh air immediately. If breathing has stopped, seek medical attention immediately; keep the person warm and at rest. Perform CPR if trained to do so.
- Do not strike vessels or fittings, as the seals could break and release high-pressure air and pieces of the fittings.

4.3.0 Fuel Oil and Flammable Liquid Piping Systems

Piping systems carry fuel from storage tanks to service tanks, and from there to an injection pump that delivers fuel to the engine combustion chamber. Fuel oil piping systems must be made of suitable materials that can withstand the operating pressures, structural stress, and chemicals to which they will be subjected. Piping, fittings, and valves must also be compatible with the products being stored. Where practical, piping should run from the tank along the shortest possible route to

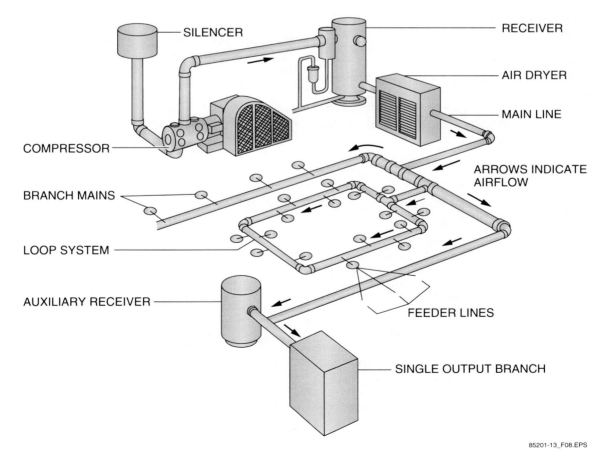

85201-13_F08.EPS

Figure 8 Compressed air piping system.

the dispensers. Piping across the tanks should be avoided whenever possible. All piping should be sloped at least $\frac{1}{8}$ inch per foot back to the tanks. When selecting materials and the type of connections to use in fuel oil piping systems, always refer to the job specifications. *Figure 9* shows a schematic diagram of a typical fuel oil piping system.

> **WARNING!**
>
> Even after the fuel tank is emptied, the vapors remaining in fuel tanks and piping can be dangerously flammable; be very cautious in working around them.

4.4.0 Steam Piping Systems

Steam is a gas that is generated by adding heat energy to water in a boiler. It is a very efficient and easily controlled heat-transfer medium. The steam piping system is used to transfer energy from a central area. Steam is commonly classified according to three ranges of pressure. The steam used in heating systems is generally at a pressure of less than 15 psi.

Steam flows naturally from a point of higher pressure, as is generated in the boiler, to a point of lower pressure in the steam mains. Because of this, the natural flow of steam does not require a pump. Steam also circulates through a heating system much faster than other types of fluids.

Condensate is the byproduct of a steam system and is formed in the distribution lines when the steam condenses to water. Once the steam condenses, the hot condensate must be removed from the system immediately and returned to the boiler. Hot steam that comes in contact with condensate that has cooled below the temperature of the steam may cause **water hammer**, which makes an annoying banging sound and can damage the pipes and fittings. It is important that there be no water in a system when the steam is introduced, because the water will expand at a high rate of speed and will damage the system. *Figure 10* shows a schematic diagram of a typical steam piping system.

4.5.0 Water Piping Systems

Among the types of water carried by piping systems are utility water, potable water, demineralized water, distilled water, and cooling water. Utility water is used primarily for cleaning, washing down, and other purposes. It is untreated water. Potable water is drinkable water. Demineralized water is treated to remove minerals that may corrode piping or damage equipment. Distilled water is a specially purified type of water. Cooling water is used as a medium for displacing heat and is also untreated water.

A wide range of piping materials can be used to convey water. The engineering specifications govern the type of materials used based on the type of water, operating pressure, surrounding conditions, and application. Utility water systems and cooling water systems usually use carbon steel or PVC pipe; potable water systems use threaded galvanized steel, copper, stainless steel, ductile iron, or PVC pipe. These systems must be sanitized, usually by being flushed and treated with chlorine, before use.

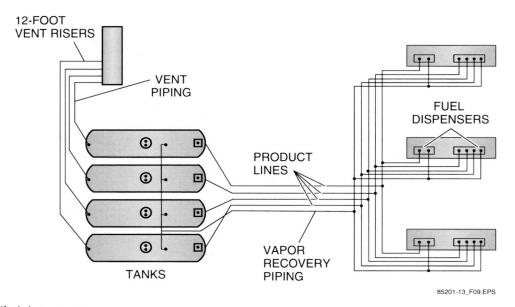

85201-13_F09.EPS

Figure 9 Fuel oil piping system.

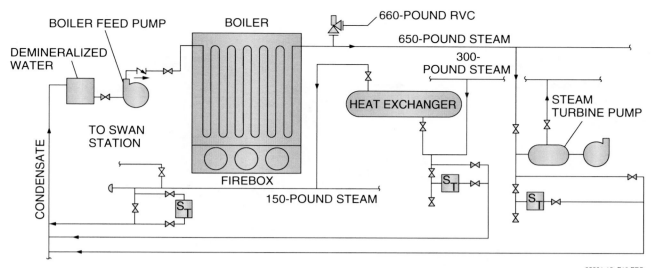

Figure 10 Steam piping system.

> **NOTE**
>
> In some states, a third-party contractor is required by code to perform the sanitizing procedures on potable water systems.

Both demineralized water systems and distilled water systems use stainless steel pipe. *Figure 11* shows a typical potable water piping system. *Figure 12* shows a typical cooling water piping system.

5.0.0 THERMAL EXPANSION OF PIPES

Metal pipes expand and contract as the temperature changes. The change can be caused by the surrounding air temperature or by the material carried within the pipe. In addition to temperature-induced changes, movement of the ship will cause pipes to stretch and bend. Unless the system is designed to account for these stresses, they could cause a piping system failure. Such failures usually occur at a joint. Expansion loops and expansion joints (*Figure 13*) are among the methods used to overcome these stresses. Proper design of the piping run can also help in overcoming thermal and physical stresses.

Under normal conditions, metal piping changes several inches in a 100' length for a change of 300°F (149°C). If only one point in a pipeline were kept fixed during a change in length, movement would take place in perfect freedom, and no stress would be imposed on the pipe or the connection. However, piping systems have more than one fixed point. They are nearly always restrained at terminal points by anchors, guides, stops, rigid hangers, or sway braces. These restraining points resist expansion and put the line under stress, which causes it to deform when the temperature changes.

Stress imposed by expansion must be kept within the strength capability of the pipe and also of the points to which the pipe is anchored. Equipment that is fastened solidly, such as a pump or a turbine, will be put under excessive stress if the pipe fastened to it expands toward it. This must be prevented by allowing expansion away from the equipment. Failure to do so can result in ruptured pipe or damage to the structure or the equipment to which the pipe is anchored.

All materials do not expand and contract at the same rate. This can cause problems when different materials are used in the same piping system. *Table 4* shows the expansion of different types of pipe.

To find out how much a length of pipe would expand for a given change in temperature, multiply the length of the pipe in inches, times the temperature change in degrees, times the coeffi-

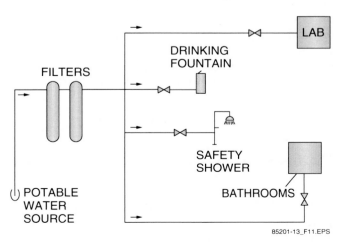

Figure 11 Potable water piping system.

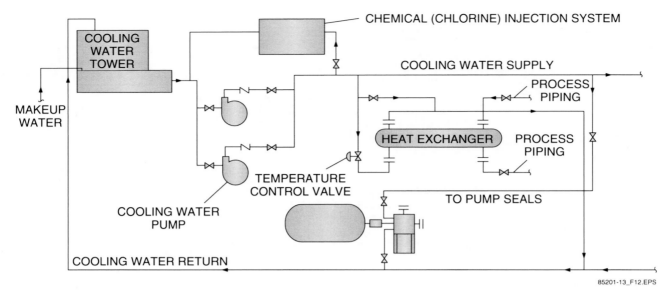

Figure 12 Cooling water piping system.

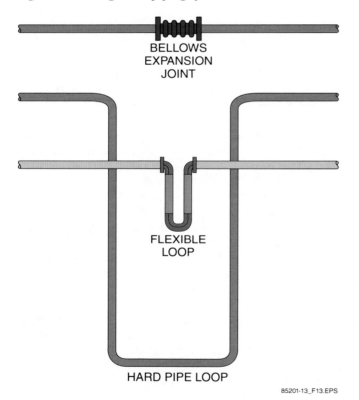

BELLOWS
EXPANSION
JOINT

FLEXIBLE
LOOP

HARD PIPE LOOP

85201-13_F13.EPS

Figure 13 Expansion joint and expansion loops.

cient of expansion for that material from *Table 4*. For example, a hundred feet of carbon steel pipe (1,200 inches) that is heated from 100°F to 500°F would expand by 1,200 (length) × 400 (temperature change) × 0.0000063 (coefficient of expansion) = 3.024 inches of expansion.

There are three accepted methods used to account for expansion and to avoid excessive stresses on the line: flexibility in the layout, installing expansion loops, and cold springing.

Table 4 Expansion of Different Types of Pipe

Coefficient of Expansion	
Material	**inches or mm**
Carbon steel	0.0000063
Aluminum	0.0000124
Cast iron	0.0000059
Nickel steel	0.0000073
Stainless steel	0.0000095
Concrete	0.0000070

85201-13_T04.EPS

5.1.0 Flexibility in Layout

Flexibility can often be built into the layout of a piping system. If a pipeline is built with small diameter pipe, and if it will not have great temperature changes, the system can be designed to account for expansion. Flexibility in layout consists of making many changes in direction between anchor points. These changes in direction allow for some thermal expansion of the line without damaging the joints and fittings. *Figure 14* shows an example of flexibility in layout of a pipeline constructed of 4-inch Schedule 40 steel pipe carrying steam indoors at 400°F (204°C).

5.2.0 Installing Expansion Loops or Expansion Joints

The second method used to account for expansion is expansion loops that are installed in the pipe run to provide extra flexibility between anchors. Expansion loops accommodate thermal expansion and contraction and sideways bowing. These loops absorb the movement of the line caused by

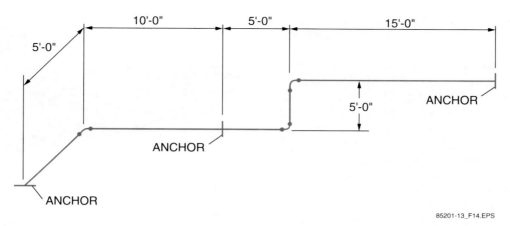

Figure 14 Flexibility in layout.

expansion and contraction. The pipe is anchored along its length and sometimes in the loop itself. Guides are placed at designated intervals along the pipe run to ensure that the pipe moves toward the expansion loop as it expands.

Expansion loops are made with pipe bends or welded pipe and fittings. They should not be fabricated with flanges or threaded fittings, because these types of joints cannot withstand the stress and will leak. Fabricated expansion U-loops made of pipe welded to long radius elbows are usually very simple in shape. They can be fabricated in the shop according to the specifications on the layout drawing. The number and types of expansion loops used differ with each installation. *Figure 15* shows an expansion loop.

The bellows-type expansion joints can be made of metal such as stainless steel, plastic, or elastomeric (rubber). In addition to compensating for thermal expansion, an expansion joint can be used to reduce noise and vibration. Piping design-

ers analyze the type of deflection (thermal, shear, tension, compression, bending) that will be to a given piping run and will select the expansion joint design suited to the conditions. It is therefore important to use only the specified expansion joints and to install them at the points shown on the drawings.

When installing expansion joints, make sure they are properly supported; expansion joints are not intended to bear weight. Rubber expansion joints may require lubrication with a graphite or glycerin lubricant to prevent them from adhering to the pipe flanges.

5.3.0 Cold Springing

The third way to deal with expansion is to use cold springing, along with one of the other two methods used to account for expansion. Cold springing is a method of shortening the pipe run when it is installed so that when the pipe heats

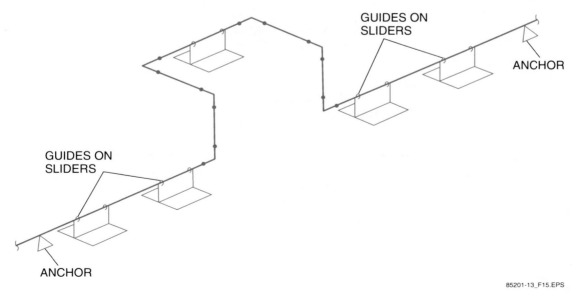

Figure 15 Expansion loop.

up, it expands to its intended length. Cold springing also relieves the stress that thermal expansion exerts on the pipe.

The cold spring gap refers to the difference between the length of the installed pipe and the length that the pipe should be. The location and the size of the cold spring gap must be calculated accurately and the piping must be installed according to those calculations. Cold springing pipe at the wrong location or by the wrong amount can damage the pipe joints, anchors, valves, or attached equipment, so it is not advisable to guess at cold spring dimensions and locations. There are several ways to implement a cold spring gap. Engineers will calculate the size of the gap. *Figure 16* shows examples of cold springing.

6.0.0 PIPE INSULATION

Piping is often covered with insulation (*Figure 17*) in order to maintain the temperature of the fluid in the pipe. Insulation is applied by craft workers who specialized in insulation. However, pipefitters should be familiar with piping insulation and its requirements. Drawings and specifications will identify which pipes are to be insulated and with what material. When installing pipe, the installers must keep in mind the space needed for insulation and ensure that there is enough clearance between pipe runs and between pipes and equipment or structures.

85201-13_F17.EPS

Figure 17 Insulated pipes.

Piping used to transport very cold materials, such as liquefied natural gas, should be insulated. Insulation helps prevent condensation on the exterior of a cold pipe as it passes through a warm space. Failure to insulate pipelines that convey steam and other hot fluids results in a tremendous loss of heat. Insulated pipelines also provide safer and more comfortable working conditions. Pipefitters must be able to recognize three types of insulation: personal protection, heat conservation, and cold conservation. Personal protection insulation provides protection to workers who may come in contact with hot pipes. Any pipe that is hotter than 110°F (43°C) must be insulated between the working level and a height of 7 feet. The working level can be considered the ground or a suspended work platform that the pipe is routed past. If a hot pipe passes by a work platform, the personal protection insulation is placed on the pipe at the working level up to 7 feet above the platform.

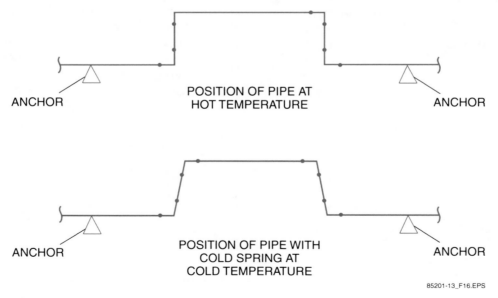

ANCHOR

POSITION OF PIPE AT
HOT TEMPERATURE

ANCHOR

ANCHOR

POSITION OF PIPE WITH
COLD SPRING AT
COLD TEMPERATURE

ANCHOR

85201-13_F16.EPS

Figure 16 Cold springing.

Heat conservation insulation is used to maintain warm temperatures within pipelines. Cold conservation insulation is used to maintain cold temperatures within pipelines. Materials used for insulation vary and are selected according to the requirements of the application. *Table 5* lists insulating materials for different temperature applications.

Mineral wool or fiberglass insulation is usually wrapped with polyethylene. As in other characteristics of piping systems, insulation is usually specified in the drawings, although it may be only by reference to a standard or specification.

Table 5 Insulating Materials for Different Temperature Applications

Heat Conservation Insulation		Cold Conservation Insulation	
Material	**Temperature Range**	**Material**	**Temperature Range**
Fiberglass	Up to 400°F	Foam rubber	Down to 0°F
85% magnesia	Up to 600°F	Mineral wool	Down to −150°F
Mineral wool	Up to 1,200°F	Rock cork	Down to −250°F
Calcium silicate	Up to 1,200°F	Rigid foam plastic	Down to −400°F

85201-13_T05.EPS

SUMMARY

The purpose of this module is to provide an overview of the types of piping systems that may be encountered by maritime pipefitters. You must always determine what a piping system contains before operating valves, inspecting for leaks, or performing maintenance on the system. Color codes can provide information on hazards and the contents of a system. Some piping systems carry flammable material and others are under pressure, so it is important to know what is carried by any piping run and to take the necessary precautions.

The majority of pipe used in maritime applications is made of mild steel. However, pipes made of other materials, including copper, stainless steel, plastic, fiberglass, and cupronickel, are used in some applications.

Expansion and contraction of piping runs is caused by temperature changes in the air surrounding the pipe, as well as the material being carried by the pipe. Expansion and contraction can also be caused by the flexing of the vessel. Proper design of the piping system allows for expansion and contraction. Expansion loops and expansion joints are used to counteract these actions.

Pipes that carry warm fluid through unheated spaces generally need to be insulated to prevent heat loss. Pipes carrying cold fluids through heated spaced may also need to be insulated. This is done in order to prevent the condensate that forms on the outside of the pipe from causing corrosion of the pipe or from dripping onto surfaces below the pipe.

1. Water that is safe to drink is called _____.
   a. utility water
   b. demineralized water
   c. potable water
   d. cooling water

2. Piping used to carry flammable gases is identified by the color _____.
   a. red
   b. yellow
   c. green
   d. bright blue

3. If a pipe is painted or marked in blue, it is probably carrying _____.
   a. seawater
   b. freshwater
   c. fuel oil
   d. steam

4. Butt welding is a method used to join _____.
   a. copper pipe
   b. plastic pipe
   c. fiberglass pipe
   d. carbon steel pipe

5. The nominal size and outside diameter of carbon steel pipe are the same from _____.
   a. 6 inches and up
   b. 8 inches and up
   c. 10 inches and up
   d. 14 inches and up

6. A pipe schedule is a method of identifying its _____.
   a. length
   b. outside diameter
   c. inside diameter
   d. wall thickness

7. The smallest schedule pipe that can be threaded is Schedule _____.
   a. 10
   b. 20
   c. 40
   d. 60

8. Only hard copper can be joined by brazing.
   a. True
   b. False

9. The type of copper pipe used for HVAC refrigerant piping is _____.
   a. Type K
   b. Type ACR
   c. Type L
   d. Type M

10. Cupronickel pipe comes in the same sizes and schedules as _____.
    a. copper pipe
    b. plastic pipe
    c. fiberglass pipe
    d. carbon steel pipe

11. The choice of piping material is critical for chemical systems because _____.
    a. the temperatures are very high
    b. chemical substances can attack some types of pipe
    c. the materials are so expensive
    d. the flow rates are extremely high

12. Piping used to carry fuel oil is _____.
    a. always made of plastic
    b. sloped back toward the fuel tank
    c. always run across the fuel tank
    d. generally painted green

13. Water hammer occurs when hot steam comes into contact with _____.
    a. condensate
    b. air
    c. pipes
    d. filters

14. The product that occurs when steam condenses to water is _____.
    a. seawater
    b. water hammer
    c. condensate
    d. bilge water

15. Which of the following is *not* a method used to counteract thermal expansion?
    a. Cold springing
    b. Insulation
    c. Flexibility in layout
    d. Expansion loops

Trade Terms Introduced in This Module

Acid: A chemical compound that reacts with and dissolves certain metals to form salts.

Butt weld: A method of joining in pipe in which the ends butt together without overlapping and are then welded using a filler metal.

Caustic: A material that is capable of burning or corroding by chemical action.

Condensate: The liquid product of steam, caused by a loss in temperature or pressure.

Cupronickel: An alloy consisting of copper and nickel with small amounts of manganese and iron.

Material safety data sheet (MSDS): A document that describes the composition, characteristics, health hazards, and physical hazards of a specific chemical. It also contains specific information about how to safely handle and store the chemical and lists any special procedures or protective equipment required. The MSDS is being replaced by the safety data sheet (SDS) as part of the GHS program.

Pounds per square inch gauge (psig): Amount of pressure in excess of the atmospheric pressure level.

Socket weld: A method of joining pipe in which the pipe end is inserted into the recessed area of a fitting or valve and then welded using a filler metal.

Water hammer: An increase in pressure in a pipeline caused by a sudden change in the flow rate. In a steam line, water hammer is caused by condensate blocking the flow of steam at a pipe bend.

Additional Resources

This module presents thorough resources for task training. The following resource material is suggested further study.

IPT Pipe Trades Handbook, 2012. Edmonton, Alberta, CA: IPT Publishing and Training.
The Pipefitters' Blue Book, 2010. W.V. Graves. Clinton, NC: Construction Trades Press.

Figure Credits

Courtesy of NOV Piping Systems, Module opener
Topaz Publications, Inc., Figures 1, 2, 3, 7

Courtesy of Mueller Industries, Figure 5

NCCER CURRICULA — USER UPDATE

NCCER makes every effort to keep its textbooks up-to-date and free of technical errors. We appreciate your help in this process. If you find an error, a typographical mistake, or an inaccuracy in NCCER's curricula, please fill out this form (or a photocopy), or complete the online form at **www.nccer.org/olf**. Be sure to include the exact module ID number, page number, a detailed description, and your recommended correction. Your input will be brought to the attention of the Authoring Team. Thank you for your assistance.

Instructors – If you have an idea for improving this textbook, or have found that additional materials were necessary to teach this module effectively, please let us know so that we may present your suggestions to the Authoring Team.

NCCER Product Development and Revision

13614 Progress Blvd., Alachua, FL 32615

Email: curriculum@nccer.org
Online: www.nccer.org/olf

❏ Trainee Guide ❏ Lesson Plans ❏ Exam ❏ PowerPoints Other _____

Craft / Level: _____ Copyright Date: _____

Module ID Number / Title: _____

Section Number(s): _____

Description: _____

_____

_____

_____

_____

Recommended Correction: _____

_____

_____

_____

_____

Your Name: _____

Address: _____

_____

Email: _____ Phone: _____

85202-13

Butt Weld Pipe Fabrication

Module Two

Trainees with successful module completions may be eligible for credentialing through NCCER's National Registry. To learn more, go to **www.nccer.org** or contact us at **1.888.622.3720**. Our website has information on the latest product releases and training, as well as online versions of our *Cornerstone* magazine and Pearson's product catalog.

Your feedback is welcome. You may email your comments to **curriculum@nccer.org**, send general comments and inquiries to **info@nccer.org**, or fill in the User Update form at the back of this module.

This information is general in nature and intended for training purposes only. Actual performance of activities described in this manual requires compliance with all applicable operating, service, maintenance, and safety procedures under the direction of qualified personnel. References in this manual to patented or proprietary devices do not constitute a recommendation of their use.

Objectives

When you have completed this module, you will be able to do the following:

1. Identify butt weld fittings.
2. Explain how to read and interpret butt weld piping drawings.
3. Describe how to properly prepare pipe ends for fit-up.
4. Explain how to determine pipe length between fittings for butt weld applications.
5. Describe the proper alignment procedures for various types of fittings.

Performance Tasks

Under the supervision of the instructor, you should be able to do the following:

1. Prepare pipe ends for fit-up.
2. Install backing rings.
3. Align pipe to both ends of various types of fittings, including flanges and ells.

Trade Terms

Align
Alloy
ASTM International
Bevel
Burn-through

Fit-up
Full-penetration weld
Lateral
Oxide
Root opening

Industry-Recognized Credentials

If you are training through an NCCER-accredited sponsor, you may be eligible for credentials from NCCER's Registry. The ID number for this module is 85202-13. Note that this module may have been used in other NCCER curricula and may apply to other level completions. Contact NCCER's Registry at 888.622.3720 or go to **www.nccer.org** for more information.

Contents

Topics to be presented in this module include:

1.0.0 Introduction 1
2.0.0 Butt Weld Fittings 1
   2.1.0 Elbows and Return Bends 1
   2.2.0 Branch Connections 1
   2.3.0 Caps 2
   2.4.0 Reducers 2
   2.5.0 Flanges 2
      2.5.1 Weld-Neck Flanges 2
      2.5.2 Lap-Joint Flanges 2
      2.5.3 Reducing Flanges 4
3.0.0 Butt Weld Piping Drawings 4
   3.1.0 Double- and Single-Line Drawings 4
   3.2.0 Isometric Drawings 4
   3.3.0 Piping Symbols 4
   3.4.0 Specifications Book 5
4.0.0 Preparing Pipe Ends for Fit-Up 5
   4.1.0 Preparing Edges 5
      4.1.1 Beveling Using Grinders 6
      4.1.2 Beveling Using Pipe Bevellers 6
      4.1.3 Beveling Using Cutters 6
      4.1.4 Thermal Beveling 6
   4.2.0 Cleaning Surfaces 6
      4.2.1 Inspecting Grinders Before Use 7
      4.2.2 Operating Grinders 9
5.0.0 Determining Pipe Lengths Between Fittings 11
   5.1.0 Calculating Takeout 11
      5.1.1 Long-Radius 90-Degree Elbows 11
      5.1.2 Short-Radius 90-Degree Elbows 11
      5.1.3 45-Degree Elbows 11
   5.2.0 Obtaining Proper Spacing Between Pipes and Fittings 13
   5.3.0 Calculating Pipe Lengths 13
      5.3.1 Center-to-Center Method 14
      5.3.2 Center-to-Face Method 14
      5.3.3 Face-to-Face Method 14
6.0.0 Selecting and Installing Backing Rings 14
7.0.0 Using and Caring for Clamps and Alignment Tools 16
   7.1.0 Angle Iron Jigs 16
   7.2.0 Shop-Made Aligning Dogs 17
   7.3.0 Cage Clamps 18
      7.3.1 Chain-Type Clamps 18
      7.3.2 Straight Pipe Welding Clamps 18

8.0.0 Performing Alignment Procedures 20
   8.1.0 Aligning Straight Pipe 21
   8.2.0 Aligning a Pipe to a 45-Degree Elbow 23
   8.3.0 Aligning a Pipe to a 90-Degree Elbow 24
   8.4.0 Squaring a 90-Degree Corner 25
   8.5.0 Aligning a Pipe to a Flange 25
   8.6.0 Aligning a Pipe to a Tee 27
   8.7.0 Fitting Butt Weld Valves 27

Figures and Tables

Figure 1   Butt-welded pipe joint 1
Figure 2   Elbows 1
Figure 3   Return bends 2
Figure 4   Branch connections 3
Figure 5   A butt-weld cap 3
Figure 6   Reducers 4
Figure 7   Example of a flange 4
Figure 8   Weld-neck flange 4
Figure 9   Lap-joint flange and stub end 5
Figure 10 Double- and single-line drawings, elevation view 5
Figure 11 Plan drawing 6
Figure 12 Isometric drawing 7
Figure 13 Butt weld fitting and valve symbols 8
Figure 14 Bevel 9
Figure 15 Pipe beveller 9
Figure 16 Oxyacetylene pipe-beveling machine 9
Figure 17 Portable grinder in use 10
Figure 18 Vertical bandsaw 10
Figure 19 Guillotine saw 11
Figure 20 General dimensions for welding fittings 12
Figure 21 Takeout of long-radius 90-degree elbow 13
Figure 22 Takeout of short-radius 90-degree elbow 13
Figure 23 Pipe spacing using welding rod 13
Figure 24 Center-to-center piping drawing 14
Figure 25 Center-to-center method 14
Figure 26 Center-to-face method 15
Figure 27 Face-to-face method 15
Figure 28 Common backing rings 16
Figure 29 Angle iron jigs 17
Figure 30 Chain-type channel-lock pliers 17
Figure 31 Notched angle iron 17
Figure 32 Sawing the bottom sides of a channel 17
Figure 33 Using a jig for alignment 17
Figure 34 Aligning dog 18
Figure 35 Cage clamps 19

Figures and Tables (continued)

Figure 36 Hydraulic clamp..20
Figure 37 Straight pipe welding clamp 20
Figure 38 Pipe alignment..21
Figure 39 Checking internal misalignment with a Hi-Lo gauge.....................22
Figure 40 Walking pipes together 23
Figure 41 Positioning of squares to align straight pipe 23
Figure 42 Pipe to 45-degree elbow alignment................................24
Figure 43 Aligning a 45-degree elbow using a level 24
Figure 44 Pipe to 90-degree elbow alignment................................24
Figure 45 Aligning a 90-degree elbow using a level 25
Figure 46 3-4-5 method...25
Figure 47 Aligning a flange horizontally 26
Figure 48 Positioning a square on a flange.....................................26
Figure 49 Adjusting flange 27
Figure 50 Pipe to tee alignment 27
Figure 51 Aligning tee using level 27
Figure 52 Preliminary valve alignment..28

Table 1   Sample Piping Specification Sheet...................................9

1.0.0 INTRODUCTION

Butt welds can be used to join pipe of any size. *Figure 1* shows an example of a butt-welded pipe joint. Butt welding is commonly used to join pipe that is 3 inches and larger in diameter, where socket welds are not appropriate. The mating ends of butt weld pipe must be properly cut and **beveled** and be in perfect alignment with each other for proper **fit-up**. This module introduces the types of fittings that are used in butt weld systems and the methods of fitting and aligning butt weld pipe. It includes instructions on interpreting butt weld drawings, preparing pipe ends for fit-up, and determining pipe length between fittings. This module also describes the tools and equipment used to **align** and fit-up pipe and describes how to verify proper alignment using a number of methods.

A butt weld can be used with any size pipe, and is the preferred method of joining larger pipe. Common wall thicknesses of pipe used in butt weld systems are Schedule 40, 80, and 160.

2.0.0 BUTT WELD FITTINGS

Fittings are used in conjunction with straight pipe to change the direction of flow, change the pipe diameter in a system, close off a line, join two straight pipes, or allow a branch line to tie into the main pipe run. Butt weld fittings are sized the same as straight pipe as far as diameter and wall thickness are concerned. End preparation is the same for fittings as it is for straight pipe.

2.1.0 Elbows and Return Bends

An elbow is a fitting that forms an angle between connecting pipes. An elbow is also known as an ell, a 90, or a 45. Elbows come in several different angles, but the angle is always assumed to be 90 degrees unless otherwise stated. The 45- and 90-degree elbows are the most common fittings used in butt weld piping and are available

in both long radius and short radius. Any other angle will likely be a special order item, or may be cut in the field from the standard 45- or 90-degree elbows. Long-radius elbows are considered standard and should always be used unless the drawings specify short-radius elbows. Reducing elbows are elbows that have different sized ends. They are used where the run changes sizes. *Figure 2* shows 90- and 45-degree long-radius elbows and a 90-degree short-radius elbow.

A return bend *Figure 3*) is a U-shaped fitting that sends the fluid back in the direction from which it came. They are also known as 180-degree returns and, like elbows, are available in long- and short-radius versions. Return bends are often used in boilers, radiators, and other systems in which the pipe must pass through the same area several times. Return bends are manufactured in various sizes and come in close, open, and wide patterns.

2.2.0 Branch Connections

Branch connections are fittings that divide the flow and send the fluid in two or more different directions. The branch connections used in butt-weld pipe fabrication include tees, **laterals**, crosses, and saddles *Figure 4*). Tees are the most widely used of the branch connections. They are used for making a 90-degree branch into the main pipe run. They can either be straight tees, meaning that all ends are the same diameter, or reducing tees, meaning that one end of the branch connection is smaller than the others. Tees are

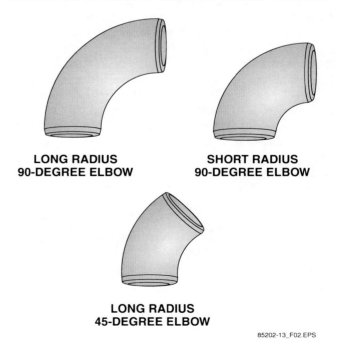

LONG RADIUS 90-DEGREE ELBOW

SHORT RADIUS 90-DEGREE ELBOW

LONG RADIUS 45-DEGREE ELBOW

85202-13_F02.EPS

Figure 2 Elbows.

85202-13_F01.EPS

Figure 1 Butt-welded pipe joint.

**LONG RADIUS
180-DEGREE RETURN**

**SHORT RADIUS
180-DEGREE RETURN**

85202-13_F03.EPS

Figure 3 Return bends.

sized by listing the dimensions straight through the tee and then listing the size of the branch, such as 3 × 3 × 2.

Laterals are another type of common branch connections that have a side opening set at any angle other than 90 degrees. Laterals can also be either straight or reducing and are sized the same way as tees. Crosses are branch connections that serve as an intersection of four pipes. They can also be either straight or reducing. Weldolets are branch connections that are commonly used to run a reduced line off a header.

A saddle is used as a reinforcement pad to a branch connection in the main pipe run. The job specifications govern the use of saddles in branch connections. The saddle can be either one piece or two pieces, with one piece split down the middle. A leakage detection hole, also known as a weep hole or tell-tale hole, is drilled into the side of the saddle. When using a split saddle, each half of the saddle contains a weep hole. If there is an internal weld failure in the branch connection beneath the saddle, the fluid within the pipe will escape through the weep hole, allowing workers to detect the problem. The weep hole can be plugged so that service can continue until a proper repair can be made.

2.3.0 Caps

Caps (*Figure 5*) are fittings that are butt-welded onto pipe ends to close off lines. Caps have the same pressure ratings as other fittings.

2.4.0 Reducers

Since straight pieces of pipe of the same size can be butt-welded together, couplings are not necessary in butt-weld systems. Reducers (*Figure 6*) are used to join pipes of different sizes in butt-weld systems in either a straight run of pipe or as part of another fitting, such as an elbow or tee. Most directional-change and branch-connection fittings are available with reduced outlet lines. Reducers are made in eccentric and concentric types. An eccentric reducer is used when the top or the bottom of the pipe must be kept level. An eccentric reducer displaces the center line of the smaller pipe at the same time that it reduces the size of the pipe. A concentric reducer is used to maintain the same center line between two pipes.

2.5.0 Flanges

A flange (*Figure 7*) is a ring-shaped plate that is attached to the end of a pipe. The flange has holes in it to allow it to be bolted to a mating flange on a pipe, fitting, or valve. Each flange has a specific pressure rating varying from 150 to 2,500 pounds. When flanges are being welded onto the end of a pipe, the face of the flange must always be protected from welding arc damage. The most common types of flanges are weld-neck, lap-joint, and reducing flanges.

2.5.1 Weld-Neck Flanges

Weld-neck flanges (*Figure 8*) are available in all pressure ratings. Weld-neck flanges provide the least turbulence and the least resistance and are often chosen for high-pressure, high-volume applications. Long weld-neck flanges are normally used for vessel and tank outlets.

2.5.2 Lap-Joint Flanges

Lap-joint flanges, also known as Van Stone flanges (*Figure 9*), are used with lap-joint stub ends. Lap-joint stub ends are straight pieces of pipe with a lap on the end that the flange rests against. These can be used at all pressures. The flange is normally not welded to the stub end, making alignment easier. When used in systems subject to high

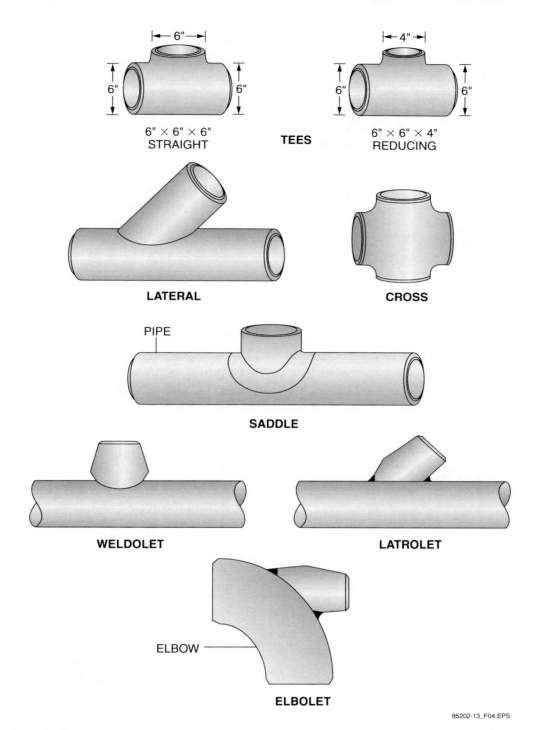

Figure 4 Branch connections.

corrosion, the flanges can usually be salvaged because they do not actually come in contact with the substance flowing through the pipe.

Lap-joint flanges are often used for economic reasons. Carbon steel lap-joint flanges can be used with stainless steel stub ends in stainless steel piping systems in order to save money. If lap-joint flanges are being welded to pipe, a lug is welded to each side of the stub end to prevent the flange from dropping when the bolts are removed.

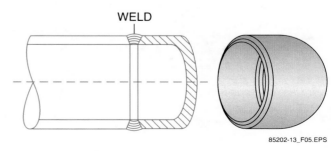

Figure 5 A butt-weld cap.

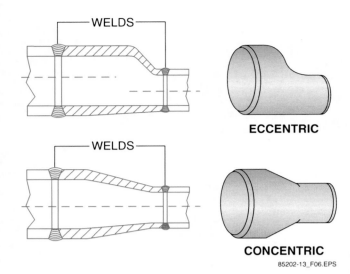

FLANGE
85202-13_F07.EPS

Figure 6 Reducers.

ECCENTRIC

CONCENTRIC

85202-13_F06.EPS

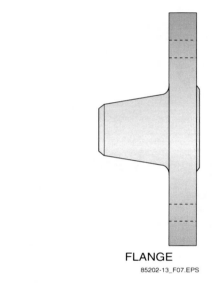

FLANGE

85202-13_F07.EPS

Figure 7 Example of a flange.

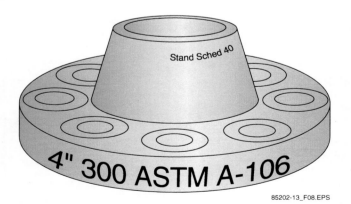

Stand Sched 40

4" 300 ASTM A-106

85202-13_F08.EPS

Figure 8 Weld-neck flange.

2.5.3 *Reducing Flanges*

Reducing flanges are used where it is necessary to change the size of a run of pipe at the flange. Otherwise, the reducing flange is attached and welded in the same manner as any other flange.

3.0.0 BUTT WELD PIPING DRAWINGS

As previously discussed, the following three types of drawings are used to represent piping systems:

- Double-line
- Single-line
- Isometric

Remember, these drawings can be either plan or elevation drawings.

3.1.0 Double- and Single-Line Drawings

Double-line drawings of a butt-weld system show the system as it actually appears, with the weld shown as a line. A single-line drawing of a butt-weld piping system shows the pipe as a single line, with a dot at the weld line. *Figure 10* is an elevation, or side view of the system. *Figure 11* shows a plan drawing, or top view, of the same system.

3.2.0 Isometric Drawings

It is common to show both the elevation and plan views of a piping system on drawings. This can be done using an isometric drawing. *Figure 12* shows an isometric drawing of the butt weld system shown in *Figures 10* and *11*, with a bill of material for the system.

3.3.0 Piping Symbols

Fittings made with butt weld joints are shown by standard symbols. However, pipefitters should not always depend on the symbol shown on the piping drawing to determine what kind of weld connection to make. Always double-check the specifications book to verify that the symbol on the drawing agrees with the specification. Symbols vary from job to job, depending on the engineering company's specifications. Always refer to the symbol legend in the piping and instrumentation drawing (P&ID) when reading piping drawings. *Figure 13* shows typical butt weld fitting and valve symbols.

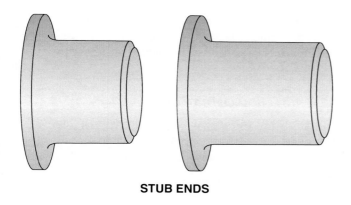

STUB ENDS

LAP-JOINT FLANGE

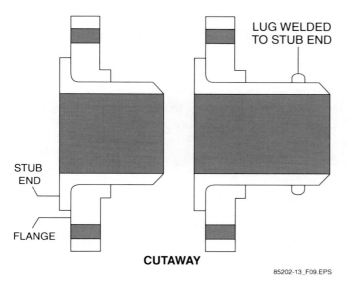

LUG WELDED
TO STUB END

STUB
END

FLANGE

CUTAWAY

85202-13_F09.EPS

Figure 9 Lap-joint flange and stub end.

3.4.0 Specifications

The specifications contain information about each pipe line. *Table 1* shows a sample page from specifications. Notice that the fittings are identified as 150# BW ASTM-A-197 CS. In this specification, the fitting is described as a 150-pound butt-weld carbon steel fitting meeting **ASTM International** Standard A-197.

4.0.0 Preparing Pipe Ends for Fit-Up

The pipe end and fitting must be prepared before a joint can be fitted up for welding. The two steps in preparing pipe ends in-

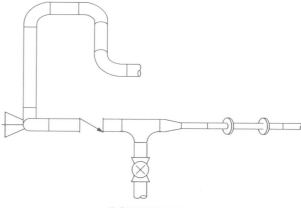

DOUBLE-LINE

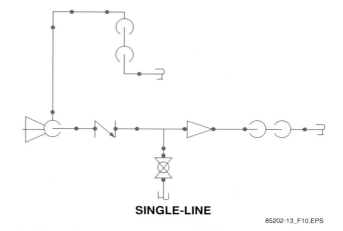

SINGLE-LINE

85202-13_F10.EPS

Figure 10 Double- and single-line drawings, elevation view.

clude preparing the edges and cleaning the surfaces.

4.1.0 Preparing Edges

Pipe edges are prepared by beveling *(Figure 14)*. A bevel is a 37.5-degree angle cut on the pipe end. The fitting may also require additional beveling. The flat edge of a pipe left after a bevel is cut is called the land. A land is often called a root face.

A beveled end with no land is called a feather edge. The land may be either a nickel edge or a dime edge. This is governed by spec. In the absence of a specification, it is normally left to the welder to decide on the land.

Pipes and fittings are often prepared in the shop and sent to the field for welding, but many times they must be prepared at the job site. Power bevellers are used to prepare pipe for butt weld fabrication. Bevellers place the correct angle, or bevel, on the end of a pipe before it is welded. There are many different ways to bevel the end of a piece of pipe to be butt-welded. The joint can be beveled mechanically, using grinders or cutters; or it can be beveled thermally, using an oxy-

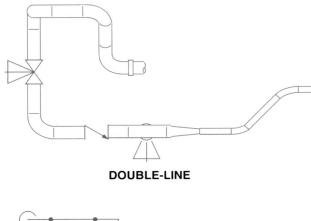

DOUBLE-LINE

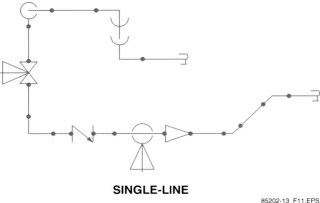

SINGLE-LINE

85202-13_F11.EPS

Figure 11 Plan drawing.

acetylene cutting torch. Mechanical joint beveling is most often used on **alloy** steel, stainless steel, and nonferrous metal piping and is frequently required for materials that could be affected by the heat of the thermal process. Mechanical beveling is slower than thermal beveling, but has the advantage of high precision with low heat and the absence of **oxides** commonly left by the thermal methods. The method used to bevel the pipe depends on the type of base material, the ease of use, and the code or procedure specifications.

4.1.1 Beveling Using Grinders

Portable hand grinders can be used to bevel pipe, but great care must be taken to make the bevel at the specified angle and keep the bevel square. Special grinders are available that can grind the bevel and land of the pipe end more precisely than a portable grinder can. These grinders are mounted inside the pipe and are locked in position to ensure that the joint preparation is square. An electric or air grinder is mounted to an arm that can be adjusted for the bevel angle required or set at 90 degrees to grind the land. The grinder is then rotated to prepare the bevel.

There are also special cutoff machines that are mounted to the outside of the pipe by a ring or

a special chain with rollers. An electric- or air-operated machine with a cutoff blade is mounted on the ring and manually or electrically powered around the pipe to cut it off.

4.1.2 Beveling Using Pipe Bevellers

The pipe beveller is a portable, air- or electric-powered beveling tool designed to bevel pipe from very small to very large diameters. The pipe beveller comes in a number of sizes, from small handheld models to very large diameter bevellers. The beveller can make precision weld preps on all metals, including stainless steel, high-alloy steel, and aluminum pipe or tubes. The beveller has a self-centering mandrel system that holds the beveller on the pipe and ensures accuracy. The beveller is one of the easiest and most accurate tools used to bevel pipe ends. *Figure 15* shows a small-diameter pipe beveller.

4.1.3 Beveling Using Cutters

Cutters have cutting tools similar to milling tools. The bevel angle is set by adjusting the cutter or changing the cutter blade. Cutters leave the surface smooth, and can also be used for cutoff operations. Another advantage of cutters is that the shape can easily be changed to prepare compound bevels. Cutters made for pipe are sometimes called pipe-end-prep lathes.

4.1.4 Thermal Beveling

Joint preparation using an oxyacetylene torch is a quick and easy way to bevel pipe ends; however, the cutting torch leaves an oxide coating on the pipe that must be ground away before welding. The oxyacetylene pipe-beveling machine (*Figure 16*) is fitted over the pipe end and tightened to the pipe. The torch is then adjusted to cut the correct angle. An electric drive motor moves the cutting torch around the pipe end to cut the preset bevel.

There are many different types of oxyacetylene pipe-beveling machines on the market. The best way to learn how to operate the pipe-beveling machine is through on-the-job training with your supervisor or the equipment manufacturer.

4.2.0 Cleaning Surfaces

After the pipe edge has been prepared, it must be cleaned before welding. Portable grinders are often used to clean the pipe edge, as shown in *Figure 17*. Grinders must be thoroughly inspected before they are used.

BILL OF MATERIALS			
P.M.	REQ'D	SIZE	DESCRIPTIONS
1		1½"	PIPE SCH/40 ASTM-A-120 GR. B
2		¾"	PIPE SCH/40 ASTM-A-120 GR. B
3	5	1½"	90° ELL ASTM-A-197 BW
4	1	1½"	TEE
5	2	¾"	45° ELL
6	1	1½" × ¾"	BELL RED. CONC.
7	2	1½"	GATE VA. BW ASTM-B62
8	1	1½"	CHECK VA. SWING BW

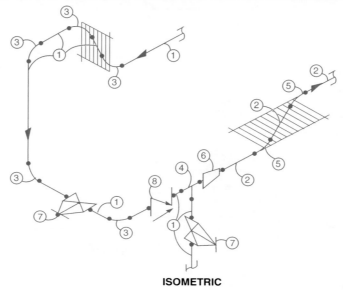

ISOMETRIC

85202-13_F12.EPS

Figure 12 Isometric drawing.

4.2.1 *Inspecting Grinders Before Use*

Whether the portable grinder being used is electric or pneumatic, a thorough inspection of the grinder is required before it is used. If the grinder is found to be defective in any way, tag the defect and do not use the grinder until the problem has been repaired. Perform the following procedures to inspect a grinder:

- Inspect the air inlet and the air line of a pneumatic grinder to ensure that there are no signs of damage that could cause a bad connection or loss of air.
- Inspect the power cord and plug on electric models to ensure that there are no signs of damage.
- Inspect the handle to ensure that it is not loose.

> **WARNING!**
>
> If the handle is loose, it could cause a loss of control.

- Inspect the grinder housing, wheel, and body for defects.
- Ensure that the trigger switch works properly and does not stick in the On position.
- Ensure that the safety guard is in good condition and securely attached to the grinder.
- Check the oil supply for pneumatic grinders.
- Ensure that the maximum rotating speed of the grinding wheel is higher than the maximum rotating speed of the grinder.
- Ensure that the grinding wheel is compatible with the metal being ground, and is tightened on the arbor.

> **CAUTION**
>
> Remember not to use grinding stones on multiple metals. It is a good idea to mark wheels to indicate what metal they have been used for.

- Start the grinder and allow it to run for one to two minutes while checking for visual abnor-

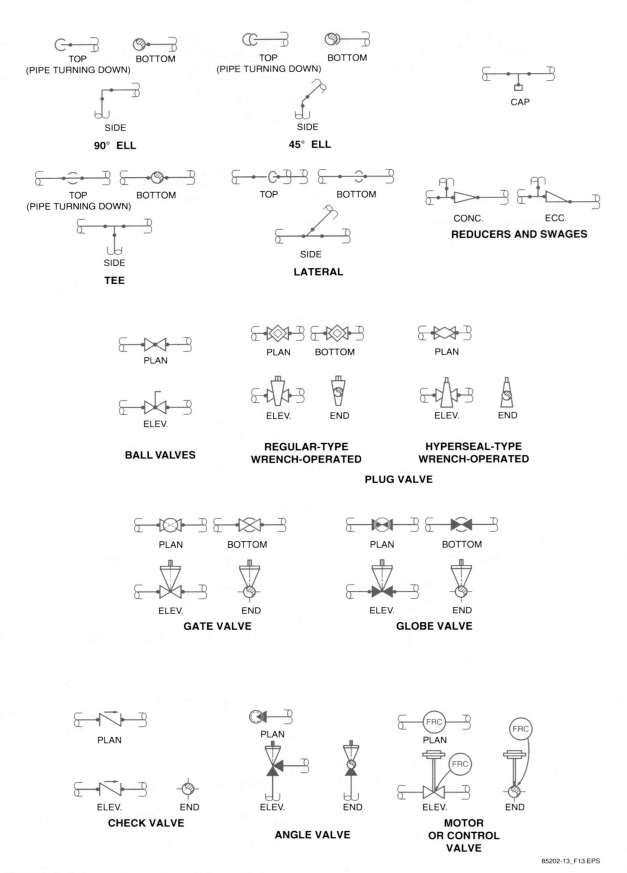

Figure 13 Butt weld fitting and valve symbols.

85202-13_F13.EPS

Table 1 Sample Piping Specification Sheet

Piping Specification	
Service: Steam	Class: ANSI 150#
Design Pressure: 150 psi	Corr. Allow: 0.030"
Max Pressure: 200 psi	Temp Limit: 200°F
Item: Size	**General Description**
Pipe: 4"	*ASTM-A-120 CS*
Fittings: 4"	150# BW *ASTM-A-197 CS*
Valves: 4"	150# Flanged *ASTM-B-105*

85202-13_T01.EPS

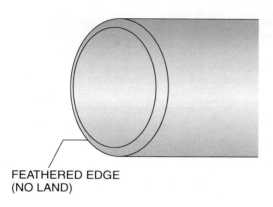

FEATHERED EDGE
(NO LAND)

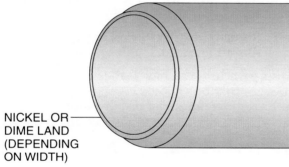

NICKEL OR
DIME LAND
(DEPENDING
ON WIDTH)

85202-13_F14.EPS

Figure 14 Bevel.

malities, excessive vibration, extreme temperature changes, or noisy operation.

• Inspect the work area to ensure the safety of yourself and others and to ensure that the heat and sparks generated by the grinder cannot start any fires.

4.2.2 Operating Grinders

When operating grinders, the operator must pay full attention to the grinder, the work being performed, and the flow of sparks and metal bits coming off the wheel. Always remember that

TOOL BLOCK

FEED HANDLES

ADAPTOR BLOCK

ID MANDREL

TOOL BIT

85202-13_F15.EPS

Figure 15 Pipe beveller.

85202-13_F16.EPS

Figure 16 Oxyacetylene pipe-beveling machine.

each grinding accessory is designed to be used in only one way. When a flat grinding disc is being used to clean a bevel, put the flat surface of the disc against the work only. Follow these steps to operate a grinder:

> **NOTE**
>
> Perform the procedures in the previous section to inspect the grinder.

Step 1 Secure the object to be ground in a vise or clamps to ensure that it does not move.

Step 2 Obtain any hot-work permits required.

Step 3 Attach the grinder to the power source.

Step 4 Position yourself with good footing and balance, and establish a firm hold on the grinder to avoid kickback.

Step 5 Pull the trigger to start the grinder.

Step 6 Apply the grinder to the work, and direct the sparks to the ground whenever possible and away from any hazards in the area, such as combustible debris or acetylene tanks.

> **CAUTION**
>
> Protect nearby alloy metals such as stainless steel tanks from cross-contamination of other metals being ground in the area.

Step 7 Apply proper force to the grinder on the grinding surface. If you are applying too much force on the grinder, you can hear the motor strain.

85202-13_F17.EPS

Figure 17 Portable grinder in use.

Step 8 Stop grinding and inspect the work and the grinding wheel periodically. If the wheel is damaged or too worn, disconnect the grinder and replace the wheel.

Step 9 Turn off and disconnect the grinder from the power source when grinding is complete.

Step 10 Inspect the grinder for any signs of damage.

Step 11 Return the grinder and any accessories to the storage area.

After grinding the bevel, the surfaces around the bevel must be cleaned. A wire brush can be used to remove any rust or corrosion. Wipe the pipe end with a clean rag to remove all grease and oil.

Stainless steel, aluminum, and other alloy pipes are cut with horizontal or vertical bandsaws

85202-13_F18.EPS

Figure 18 Vertical bandsaw.

(*Figure 18*), abrasive saws, or circular saws. The circular saws used to cut stainless steel or carbon steel are called coldsaws, because they do not produce really high temperatures while cutting, as do the abrasive cutters. The blades on coldsaws are tipped with carbide or a kind of cement. Another type of coldsaw is the guillotine saw (*Figure 19*), basically a powered bow saw, usually with hydraulic feed mechanisms. The guillotine is mounted on the pipe itself, rather than on a table.

5.0.0 DETERMINING PIPE LENGTHS BETWEEN FITTINGS

To determine the length of a piece of pipe between two butt weld fittings, the pipefitter must be able to determine the takeout of the fitting and the required welding gap, or **root opening**. Both the takeout of the fitting and the welding gap must be subtracted from the dimension of the piping run.

5.1.0 Calculating Takeout

The takeout of a butt weld fitting refers to the laying length of the fitting, or the distance from the face of one side of the fitting to the center of the fitting. The takeout of common welding fittings can be found in charts showing general dimensions for welding fittings (*Figure 20*).

The general dimensions for welding pipe fittings are supplied by fitting manufacturers and are also found in many pipefitting guidebooks.

Figure 19 Guillotine saw.

However, it is often necessary to calculate takeout in the field. Formulas are provided for the following types of fittings:

- Long-radius 90-degree elbows
- Short-radius 90-degree elbows
- 45-degree elbows

5.1.1 Long-Radius 90-Degree Elbows

The takeout of long-radius 90-degree mild steel elbows is always $1\frac{1}{2}$ times the nominal size of the elbow. This is true of all long radius 90-degree mild steel elbows except for $\frac{1}{2}$-inch elbows. For example, to determine the takeout of an 8-inch LR 90, multiply 8 times $1\frac{1}{2}$, and the answer, 12, is the takeout (*Figure 21*). Note, however, that this method does not work for non-ferrous metals.

NOTE

The takeout of an extra long radius fitting is three times the diameter.

5.1.2 Short-Radius 90-Degree Elbows

The takeout of a short-radius (SR) 90-degree elbow is always the same as the nominal size of the elbow. For example, a 3-inch SR 90 has a takeout of 3 inches, and a 6-inch SR 90 has a takeout of 6 inches (*Figure 22*).

5.1.3 45-Degree Elbows

To calculate the takeout of a 45-degree elbow, multiply the nominal size of the elbow by $\frac{5}{8}$ or 0.625. Do not divide the takeout of a 90-degree elbow by 2 to find the takeout of a 45-degree elbow of the same nominal size. It will not work.

If a calculator is not available, the takeout of a 45-degree elbow can be determined using the following method.

Step 1 Write down the nominal size of the elbow.

Example: 8 inches

Step 2 Divide this number by 2, and write the answer next to the nominal size.

Example: 8 inches 4 inches

Step 3 Divide this number by 2, and write this number next to the last number.

Example: 8 inches 4 inches 2 inches

Step 4 Divide this number by 2, and write this number next to the last number.

Example: 8 inches 4 inches 2 inches 1 inch

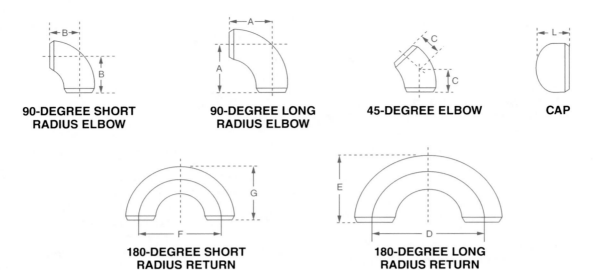

90-DEGREE SHORT RADIUS ELBOW

90-DEGREE LONG RADIUS ELBOW

45-DEGREE ELBOW

CAP

180-DEGREE SHORT RADIUS RETURN

180-DEGREE LONG RADIUS RETURN

DIMENSIONS (in inches)												
	90° Elbow		90° Elbow	180° Return Elbows				Cap	Wall Thickness			
				Long Radius		Short Radius						
Pipe Size	Long Radius Center to End A	Short Radius Center to End B	Center to End C	Center to Center D	Back to Face E	Center to Center F	Back to Face G	Length L	S	XS	Sched. 160	Pipe Size
½	1½		⅝	3	1⅞			1	0.109			½
¾	1⅛		⁷/₁₆	2½	1¹¹/₁₆			1	0.113	0.154		¾
1	1½	1	⅞	3	2³/₁₆	2	1⅝	1½	0.133	0.179	0.250	1
1¼	1⅞	1¼	1	3¾	2¾	2½	2¹/₁₆	1½	0.140	0.191	0.250	1¼
1½	2¼	1½	1⅛	4½	3¼	3	2⁷/₁₆	1½	0.145	0.200	0.281	1½
2	3	2	1⅜	6	4³/₁₆	4	3³/₁₆	1½	0.154	0.218	0.343	2
2½	3¾	2½	1¾	7½	5³/₁₆	5	3¹⁵/₁₆	1½	0.203	0.276	0.375	2½
3	4½	3	2	9	6¼	6	4¾	2	0.216	0.300	0.438	3
3½	5¼	3½	2¼	10½	7¼	7	5½	2½	0.226	0.318		3½
4	6	4	2½	12	8¼	8	6¼	2½	0.237	0.337	0.531	4
5	7½	5	3⅛	15	10⁵/₁₆	10	7¾	3	0.258	0.375	0.625	5
6	9	6	3¾	18	12⁵/₁₆	12	9⁵/₁₆	3½	0.280	0.432	0.718	6
8	12	8	5	24	16⁵/₁₆	16	12⁵/₁₆	4	0.322	0.500	0.906	8
10	15	10	6¼	30	20⅜	20	15⅜	5	0.365	0.500	1.125	10
12	18	12	7½	36	24⅜	24	18⅜	6	0.375	0.500	1.312	12
14	21	14	8¾	42	28	28	21	6½	0.375	0.500	1.406	14
16	24	16	10	48	32	32	24	7	0.375	0.500	1.593	16
18	27	18	11¼	54	36	36	27	8	0.375	0.500		18
20	30	20	12½	60	40	40	30	9	0.375	0.500		20
22	33	22	13½	66	44	44	33	10	0.375	0.500		22
24	36	24	15	72	48	48	36	10½	0.375	0.500		24

85202-13_F20.EPS

Figure 20 General dimensions for welding fittings.

Step 5 Add the second and fourth numbers that you have written down to find the takeout of the 45-degree elbow.

Example: 4 inches + 1 inch = 5 inches

Therefore, the takeout for an 8-inch, 45-degree elbow is 5 inches.

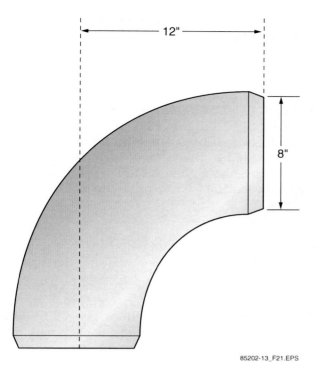

Figure 21 Takeout of long-radius 90-degree elbow.

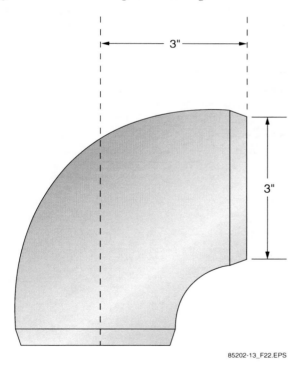

Figure 22 Takeout of short-radius 90-degree elbow.

5.2.0 Obtaining Proper Spacing Between Pipes and Fittings

To ensure that the welder has enough space to make a full-penetration weld, the pipefitter must leave a space between the pipe edges to be joined. This space is called the root opening, or gap. The proper gap is crucial for a good fit-up. If the gap is too small, it is difficult to fuse the pipes properly. If the gap is too large, burn-through may result. Always check the job specifications to determine pipe spacing. Almost all pipe-welding gaps are between $\frac{1}{16}$ and $\frac{1}{8}$ inch; however, this is determined by the job specification. A simple way to obtain the correct gap between pipes is to bend a welding rod that is the same size as the required gap and place it between the pipes as shown in *Figure 23*. A scrap piece of steel that is the correct thickness can also be used.

5.3.0 Calculating Pipe Lengths

Piping drawings usually show the dimensions of the overall run of pipe, but it is up to the pipefitter to determine the lengths of the individual straight pipes within the run. The three methods of determining pipe lengths are the center-to-center method, the center-to-face method, and the face-to-face method.

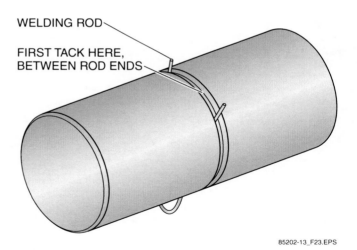

WELDING ROD

FIRST TACK HERE,
BETWEEN ROD ENDS

Figure 23 Pipe spacing using welding rod.

5.3.1 Center-to-Center Method

Figure 24 shows a piping drawing showing the center-to-center dimension. When using the center-to-center method, measurements are taken from the center of one fitting to the center of the next fitting (*Figure 25*). The actual length of pipe between butt weld fittings is determined by subtracting the takeout of each of the fittings and the required gap for each weld on each fitting.

In the center-to-center method for butt welding, the welding gap is added to the takeout of the two fittings. The sum is then subtracted from the center-to-center dimension to determine the length of pipe to be cut.

5.3.2 Center-to-Face Method

In butt weld piping, the length of a run of pipe from a fitting to a flange is determined by the center-to-face method (*Figure 26*). In the center-to-face method, the welding gap, which is normally $\frac{1}{8}$ inch, is added to the takeout of each of the two fittings. The sum is then subtracted from the cen-

ter-to-face dimension to determine the length of pipe to be cut.

5.3.3 Face-to-Face Method

In butt weld piping, the face-to-face method (*Figure 27*) is used to determine the length of pipe between two flanges. Like the center-to-face method, the welding gap, which is normally $\frac{1}{8}$ inch, is added to the takeout of each of the two fittings. The sum is then subtracted from the face-to-face dimension to determine the length of pipe to be cut.

6.0.0 SELECTING AND INSTALLING BACKING RINGS

Backing rings are generally required in piping systems where severe conditions are anticipated. Job specifications govern the use of backing rings. Other names for backing rings are chill rings and welding rings. Backing rings help in obtaining a close-tolerance fit-up. They also help to eliminate weld defects that often form on the inside of the pipe as a direct result of the welding process and that can restrict the flow of material through a system. To ensure a smooth flow across the backing ring, the ends of the ring are usually beveled about 15 degrees.

Backing rings are usually stocked in diameters ranging from $\frac{3}{4}$ inch to 36 inches, with larger sizes available on request. There are four common types of backing rings: Type CCC, Type CC, Type C, and plain rings. *Figure 28* shows the four common types, along with an example of a backing

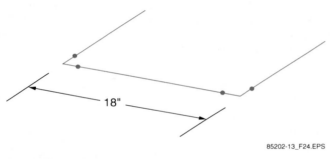

85202-13_F24.EPS

Figure 24 Center-to-center piping drawing.

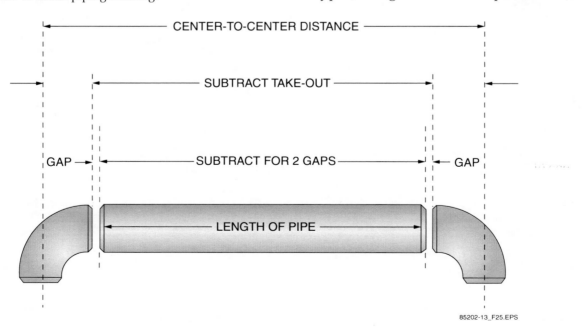

85202-13_F25.EPS

Figure 25 Center-to-center method.

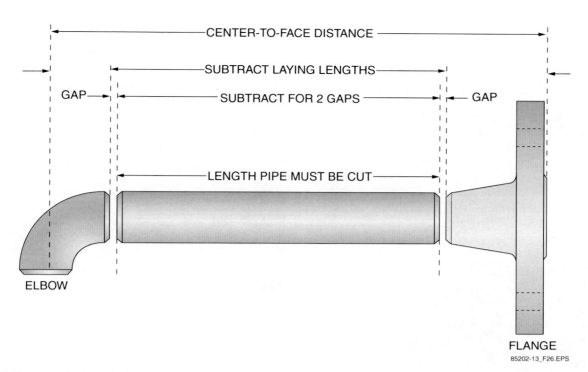

Figure 26 Center-to-face method.

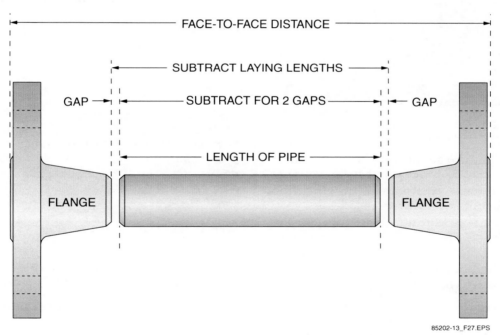

Figure 27 Face-to-face method.

ring installed in a pipe. All but the plain backing rings have nubs that are used to establish the root opening. Type CCC has the longest nubs. These nubs are knocked off with a chipping hammer after the joint is tack-welded. Type CC nubs can be knocked off or left in place to be melted during the root pass.

The Type CCC backing rings have long nubs that serve as spacers at the joint. Type CCC rings are typically used when the high-low misalign-ment between the pipe and fitting is great. The nubs are usually spot-welded onto the ring during manufacture and can be broken off cleanly with a chipping hammer once the joint is tack-welded. The nubs of the Type CC backing rings are shorter than those on the Type CCC backing rings. These rings are typically used with automatic welding. Type CC rings can either be chipped off prior to welding or they can be fused with the welding root pass. The nubs on the Type C welding ring

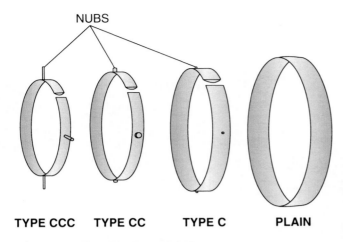

TYPE CCC **TYPE CC** **TYPE C** **PLAIN**

NUBS

85202-13_F28.EPS

Figure 28 Common backing rings.

are punched. They are actually a part of the backing ring and cannot be removed prior to welding. These rings are used in situations in which the high-low alignment is very close.

Plain backing rings do not have nubs, so the gap between the pipe and fitting must be set manually. It is important that the backing ring and the pipe being welded be made of the same material. Follow these steps to install backing rings:

Step 1 Ensure that the pipe ends have been cut square and are properly beveled with a smooth land on the face of the bevel.

Step 2 Compress the backing ring slightly, and insert it into the pipe end until the nubs are flush against the bevel land.

Step 3 Place the other pipe or fitting over the ring and butt it against the nubs. The pipe is now ready to be aligned and welded.

7.0.0 USING AND CARING FOR CLAMPS AND ALIGNMENT TOOLS

In butt weld piping, alignment is a very critical task. The pipe must be aligned with other pipes, or with fittings. A set of specialized clamps and tools for alignment has been developed. Some are clamps to hold the pieces; some are capable of aligning the pieces as well.

> **NOTE**
>
> An EB ring (named for General Dynamics Electric Boat division) is a consumable insert that is used in place of a backing ring. It is commonly used when welding pipes on nuclear-powered ships where 100 percent X-ray weld inspection is required. EB rings are used for pipe ranging from ⅛" to 24".

7.1.0 Angle Iron Jigs

Angle iron jigs are fabricated on the job site and are used to hold smaller diameter pipe for welding. They can be fabricated out of angle iron or channel iron that will hold the pipe in line or at the desired angle for welding. *Figure 29* shows some typical angle iron jigs. Chain-type channel-lock pliers (*Figure 30*) can be used to secure the pipe to the angle iron.

Follow these steps to fabricate an angle iron jig:

> **NOTE**
>
> This procedure explains how to make an angle iron jig from ⅛- by 1½-inch channel iron that is 3 feet, 9 inches long. Channel iron of this size can be used for pipe sizes ranging from 1¼ inch to 3 inches.

Step 1 Cut out 90-degree notches about 9 inches from one end (*Figure 31*).

Step 2 Using a torch, heat the bottom of the notch until the metal is red-hot.

Step 3 Bend the channel iron into a 90-degree angle.

Step 4 Have the welder weld the sides of the corners.

Step 5 Place an elbow in the jig and saw halfway through the sides of the channel at the bottom face of the elbow (*Figure 32*).

Step 6 Repeat Step 5 with several different sizes of elbows so that the jig can be used for several applications. A used hacksaw blade can be placed in the notch to provide the proper welding gap (*Figure 33*).

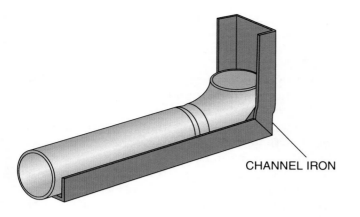

CHANNEL IRON

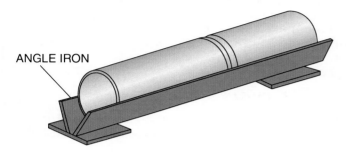

ANGLE IRON

CHANNEL IRON

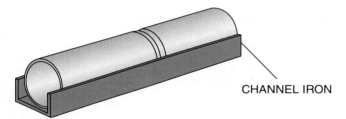

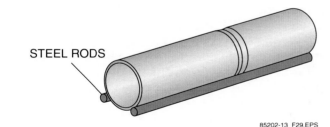

STEEL RODS

85202-13_F29.EPS

Figure 29 Angle iron jigs.

85202-13_F30.EPS

Figure 30 Chain-type channel-lock pliers.

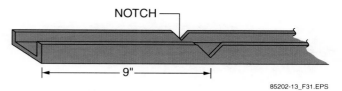

NOTCH

9"

85202-13_F31.EPS

Figure 31 Notched angle iron.

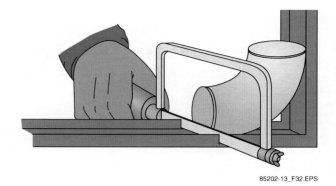

85202-13_F32.EPS

Figure 32 Sawing the bottom sides of a channel.

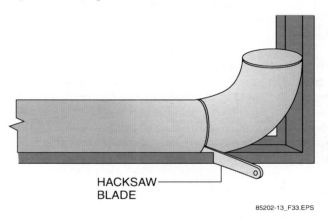

HACKSAW BLADE

85202-13_F33.EPS

Figure 33 Using a hacksaw blade for alignment.

7.2.0 Shop-Made Aligning Dogs

Many pipefitters fabricate their own aligning devices. In the pipefitting trade, these devices are known as aligning dogs. *Figure 34* shows an example of an aligning dog.

> **NOTE**
>
> Aligning dogs are not permitted under some codes.

The aligning dog shown in *Figure 34* is made from plate of the same material as the pipe being welded, to which a nut has been welded. The aligning devices are tack-welded onto one section of the pipe so that the other section can be moved by tightening the bolt. Usually, two such dogs would be used to provide more control over the fit-up. Once the fit-up is complete, the aligning dogs are removed from the pipe. After the dog is

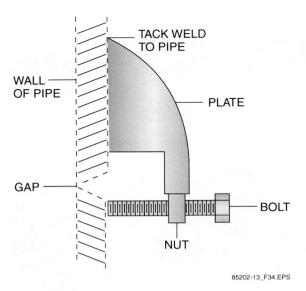

Figure 34 Aligning dog.

removed from the pipe, the pipe must be ground to remove any welding deposits. Aligning dogs also require periodic grinding to remove welding deposits.

Welding codes in some areas prohibit anything being welded onto the pipe. Therefore, an aligning dog such as the one shown would not be permitted. Make sure you know the code before using any aids that must be welded to the pipe. If codes do not permit the dog to be tacked on, make three dogs with the weld leg parallel top and bottom. Three of these dogs can be used to align the pipe by placing them at three points 120-degrees apart on the pipe and holding the weld leg against the pipe with a chain and a ratchet tensioner or chain-type clamping pliers of the type previously shown.

> **NOTE**
> Another common type of pipe alignment clamp, called the ultra-clamp, is a screw-type clamp that allows the alignment of the joint to be adjusted with the adjusting screws. It makes contact with the two pieces at only three points, and is made in three sizes, 1 to 2 inches, 2 to 6 inches, and 5 to 12 inches.

7.3.0 Cage Clamps

Lever-type cage clamps are very useful if several straight butt joints must be made in a single piping run. They are usually made for one size of pipe, ranging from 2 to 36 inches in diameter. Larger sizes are available through special manufacturers. These clamps cannot be used to align flanges.

Cage clamps are usually made from two rings joined by four or more horizontal bars. In the mechanical type, a lever is used to open and close the clamp. When the lever is drawn tight, the horizontal bars butt against the pipe, aligning each section with the other. Fine adjustments are usually possible by loosening or tightening lock nuts attached to the arms that are controlled by the lever (*Figure 35*). Locking-screw clamps are adjusted using a ratchet.

Hydraulic cage clamps (*Figure 36*) are similar to lever-type cage clamps. They are used on larger pipe ranging from 16 to 36 inches in diameter, with larger sizes manufactured through special order. The hydraulic cage clamp operates in the same manner as the lever-type clamp, except it uses a hydraulic jacking mechanism to tighten the clamp against the pipe.

7.3.1 Chain-Type Clamps

Chain-type welding clamps use the same principle as angle iron jigs, but have chains that hold the pipe in place. These clamps are made of tough, durable metal that will not be warped by the heat of the welding. Chain-type clamps are generally available for the common fit-ups, such as straight pipe, elbow, T-joint, and flange fit-ups.

7.3.2 Straight Pipe Welding Clamps

Straight pipe clamps are chain-type clamps used to secure two straight pipes together end to end. They consist of straight sections of steel approximately 14 to 18 inches long and two lengths of chain, one near each end, attached to screw locks that tighten the chain once it has been wrapped around the pipe. Straight pipe welding clamps (*Figure 37*) can be used on pipe with ½- to 8-inch outside diameters.

Follow these steps to use and care for a straight pipe welding clamp:

> **NOTE**
> For this exercise, assume that two previously prepared 4-inch steel pipes must be welded together end to end.

Step 1 Inspect the straight pipe clamp for any obvious damage, such as broken or worn chain links, bent or broken clamp feet, a warped clamp back, bent or broken chain hooks, damaged chain tighteners, and grease, rust, or excessive dirt. Repair or replace any damaged part. Clean and oil any part as needed.

CAGE CLAMP ON PIPE

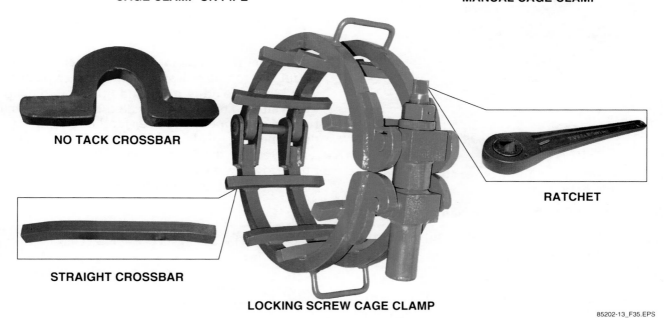

MANUAL CAGE CLAMP

NO TACK CROSSBAR

STRAIGHT CROSSBAR

LOCKING SCREW CAGE CLAMP

RATCHET

85202-13_F35.EPS

Figure 35 Cage clamps.

Step 2 Place the pipes on the clamp body so that the pipes line up end to end. Install a backing ring if it is required.

Step 3 Set the proper gap between the pipes.

Step 4 Position the clamp on top of, and evenly spaced over, the two sections of pipe so that the chains drop down the backside of the clamp as you face the pipes.

Step 5 Hold the clamp in place with one hand, and reach under the pipe and grasp the chain with the other hand.

Step 6 Pull the chain around the pipe and up to the locking notch on the clamp.

Step 7 Secure the chain to the clamp.

Step 8 Rotate the clamp around the pipe to the bottom of the pipe.

Step 9 Turn the screw handle to tighten the chain. Tighten only until the chain is snug. Some readjustment may be needed later.

Step 10 Check the alignment of the unclamped section of pipe, to be certain that it is still

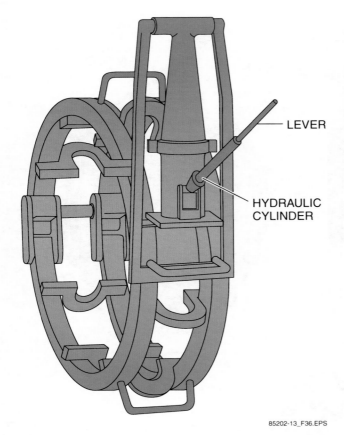

Figure 36 Hydraulic clamp.

85202-13_F36.EPS

LEVER

HYDRAULIC CYLINDER

aligned with the first section of pipe and is ready to be welded.

Step 11 Wrap the chain around the second section of pipe.

Step 12 Attach the chain to the clamp.

Step 13 Tighten the second chain onto the second section of pipe.

Step 14 Adjust both chains until they are tight.

Step 15 Check the pipe sections again to ensure that they are still aligned with each other and ready for welding.

Step 16 When the welding has been done, loosen the chains and remove the clamp.

WARNING!

Handle the pipe and the clamps with care. Welded objects remain hot for several minutes after the welding is completed.

Step 17 Clean the clamp, if necessary, and inspect it for damage. If it is damaged, replace it.

Step 18 Store the straight pipe clamp.

85202-13_F37.EPS

Figure 37 Straight pipe welding clamp.

8.0.0 PERFORMING ALIGNMENT PROCEDURES

Aligning pipe for welding requires great skill. The pipes must be correctly aligned to make a smooth inner wall that does not restrict flow. The pipes must also be aligned to put the completed pipe in the proper position with respect to other pipes and fittings in the piping system. This is complicated by the fact that fittings and pipe may have small but significant differences in inside diameter. If this is detected, it may require the fitter to counterbore the pipe or fitting to allow a closer match. The joint cannot be offset or angled, but must be joined evenly (*Figure 38*).

NOTE

During the alignment of a fitting, compensate for movement of pipe due to weld draw. This should be taken into account prior to weld completion.

If the walls of the pipe or fittings being joined are not exactly the same thickness, ensure that the internal misalignment is the same all the way around the joining pipes. It is also necessary to ensure that the internal misalignment is less than the maximum allowable tolerance as stated in the specifications. If there is internal misalignment on one side only, either the pipe or the fitting may be out of round and therefore must be replaced. This happens most often in large-diameter pipe.

Pipes can be aligned using a level, squares, or a Hi-Lo gauge. Levels and squares are used to check end alignment; the primary purpose of a Hi-Lo gauge is to check for inner wall misalignment. The name of the gauge comes from the

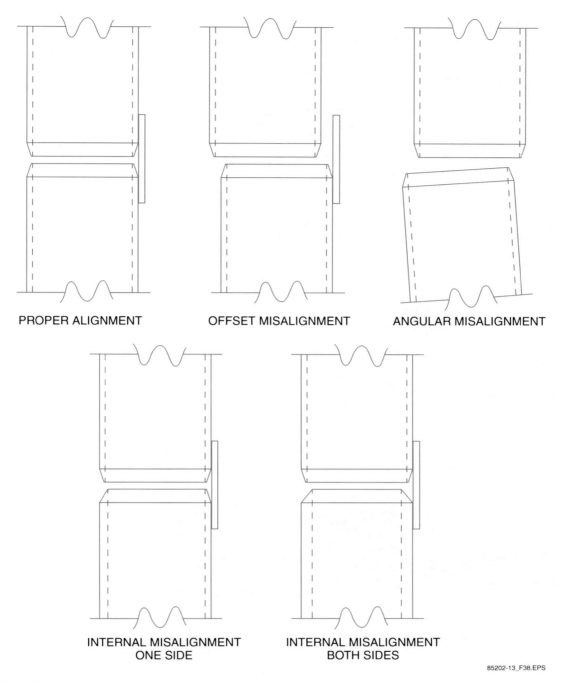

PROPER ALIGNMENT OFFSET MISALIGNMENT ANGULAR MISALIGNMENT

INTERNAL MISALIGNMENT
ONE SIDE

INTERNAL MISALIGNMENT
BOTH SIDES

85202-13_F38.EPS

Figure 38 Pipe alignment.

relationship between the inside wall of one pipe and that of another pipe. To check for internal misalignment, Hi-Lo gauges have two prongs, or alignment stops, that are pulled tightly against the inside wall of the joint so that one stop is flush with each side of the joint. The variation between the two stops is read on a scale marked on the gauge. To measure misalignment using a Hi-Lo gauge, insert the prongs of the gauge into the joint gap. Pull up on the gauge until the prongs are snug against both inside surfaces, and read the misalignment on the end of the scale. *Figure*

39 shows internal misalignment being checked with a Hi-Lo gauge. Major misalignment could be predetermined by measuring pipe circumference before fit-up.

8.1.0 Aligning Straight Pipe

Straight pipe is aligned using framing squares. Two framing squares are needed to align two pieces of straight pipe. If the pipe being joined is less than 10-inch nominal size, support the pipe with three-legged jack stands. If the pipe is larger

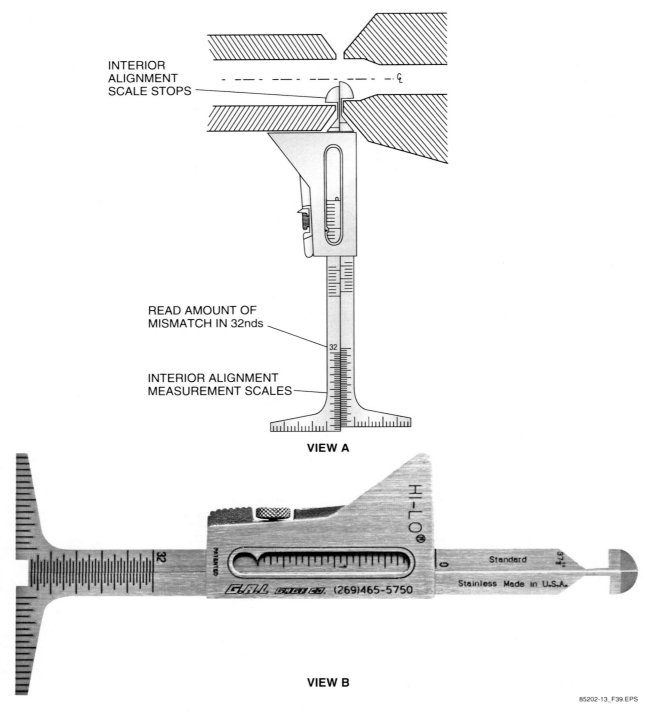

INTERIOR ALIGNMENT SCALE STOPS

READ AMOUNT OF MISMATCH IN 32nds

INTERIOR ALIGNMENT MEASUREMENT SCALES

VIEW A

VIEW B

85202-13_F39.EPS

Figure 39 Checking internal misalignment with a Hi-Lo gauge.

than 10-inch nominal size, use four-legged jack stands. Follow these steps to align straight pipe:

Step 1 Set up four pipe supports in a straight line to hold the two pipes.

> **NOTE**
>
> For this exercise, assume that two 6-foot sections of 4-inch steel pipe must be welded into one 12-foot section.

Step 2 Place the pipes on the supports so that the pipes line up end to end.

Step 3 Set the proper gap between the pipes. When setting the gap between larger sizes of pipe, the jack stands can be tapped with a hammer opposite to the direction that you are trying to move the pipe. This will walk the pipes together from a maximum distance of about 3 inches (*Figure 40*).

Step 4 Place the framing squares on the top centers of the pipes and slide the squares together. *Figure 41* shows how to position the squares.

Step 5 Adjust the second pipe up or down until the long legs of the two squares match up at every point.

Step 6 Remove the squares and have the welder tack-weld the coupling to the pipe at the 12 o'clock and 6 o'clock positions. Remove the spacer after the first tack weld.

> **NOTE**
>
> Step back and recheck alignment after each tack weld. Make sure the welder is exactly sure where to place each tack weld.

Step 7 Rotate the pipe 90 degrees in the jack stands, and repeat the alignment with the squares. Have the welder tack-weld the coupling to the pipe at the new 12 o'clock and 6 o'clock positions.

> **WARNING!**
>
> Handle the pipe with care. Welded objects remain hot for several minutes after the welding is completed.

8.2.0 Aligning a Pipe to a 45-Degree Elbow

Two framing squares can be used to align a 45-degree elbow to the end of a straight pipe when using smaller diameter pipe. Follow these steps to align straight pipe to a 45-degree elbow:

Step 1 Set up a 6-foot section of 4-inch steel pipe on two pipe supports.

Step 2 Set up a pipe jack at the end of the pipe to support the elbow.

Step 3 Place the 45-degree elbow on the support so that it lines up with the pipe and turns to the side. Install a backing ring between the pipe and the elbow if required.

Step 4 Set the proper gap between the pipe and the elbow.

Step 5 Hold the elbow in line with the pipe to make a preliminary alignment.

Step 6 Have the welder make a light tack-weld at the 12 o'clock position. Since the elbow is turned to one side, this tack is actually made on the side of the elbow.

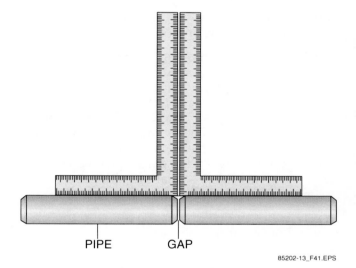

PIPE   GAP

85202-13_F41.EPS

Figure 41 Positioning of squares to align straight pipe.

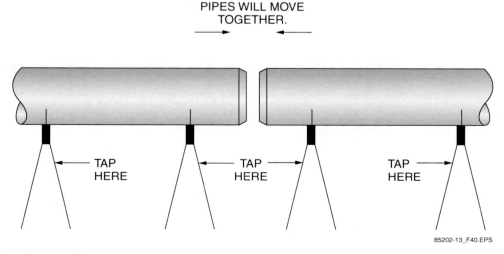

PIPES WILL MOVE TOGETHER.

TAP HERE   TAP HERE   TAP HERE

85202-13_F40.EPS

Figure 40 Walking pipes together.

Step 7 Rotate the pipe and elbow 180 degrees so that the elbow faces the opposite direction.

Step 8 Set the proper gap between the pipe and the fitting.

Step 9 Have the welder make a light tack-weld at the 12-o'clock position.

Step 10 Rotate the pipe and elbow 90 degrees so that the elbow turns up.

Step 11 Place the tongue of one framing square on the top center of the straight pipe so that the body is vertically in line with the end of the pipe.

Step 12 Place the body of the second framing square on the end of the elbow.

Step 13 Adjust the elbow so that the same inch marks on the tongue and the body of the second square contact the inner edge of the first square (*Figure 42*). When fabricating larger diameter pipe, the framing square is not large enough to use with the elbow. In this case, level the pipe in the jack stands, and use a 24-inch spirit level with a 45-degree vial to set the position of the elbow (*Figure 43*).

Step 14 Have the welder tack-weld the top and bottom of the joint.

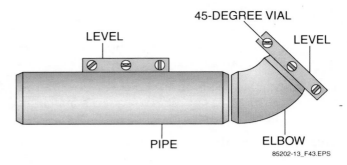

Figure 43 Aligning a 45-degree elbow using a level.

Step 15 Check the alignment both ways and adjust it if necessary by tapping the fitting with a hammer.

Step 16 Have the welder tack-weld the sides of the joint.

8.3.0 Aligning a Pipe to a 90-Degree Elbow

Two framing squares can also be used to align a 90-degree elbow to the end of a straight pipe. Prepare and put the first tack-weld on the pipe and elbow at the 12 o'clock position on the pipe.

Step 1 Place the tongue of one framing square on the top center of the straight pipe so that the body is vertically in line with the end of the pipe.

Step 2 Place the body of the second framing square on the end of the elbow so that the tongue of the square lines up with the body of the other square (*Figure 44*).

Step 3 Adjust the elbow until the squares are flush against each other. When fabricating larger diameter pipe, the framing square is not large enough to use with the elbow. In this case, level the pipe in the

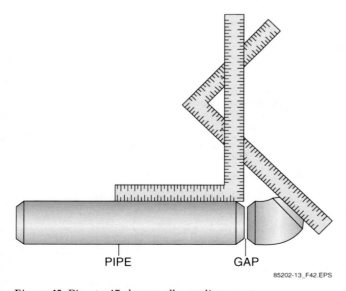

Figure 42 Pipe to 45-degree elbow alignment.

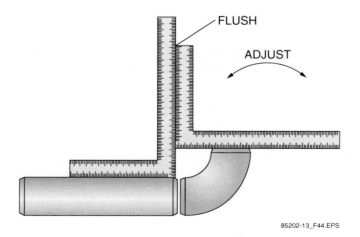

Figure 44 Pipe to 90-degree elbow alignment.

jack stands and use a 24-inch spirit level to set the position of the elbow (*Figure 45*).

Step 4 Have the welder tack-weld the bottom of the joint.

> **NOTE**
> Step back and recheck alignment after each tack weld. Make sure the welder is exactly sure where to place each tack weld.

Step 5 Place a square on the sides of the elbow and pipe and check for misalignment from side to side.

Step 6 Have the welder tack-weld the sides of the joint.

8.4.0 Squaring a 90-Degree Angle

Long pipe runs with a 90-degree angle are difficult to square with a framing square because the squares are too small. These long runs can be squared using the 3-4-5 method. The 3-4-5 method, based on the Pythagorean theorem, uses a triangle with a 3-foot side, a 4-foot side, and a 5-foot side. A triangle with these sides will always have a 90-degree angle at the intersection of the 3- and 4-foot sides. For longer runs of pipe, any multiple of 3-4-5 can be used to make a square corner. For example, the triangle can have sides of 6, 8, and 10 feet; 9, 12, and 15 feet; or 12, 16, and 20 feet. This section explains how to square a pipe into the 90-degree elbow that was fit up in the previous section of this module. The pipe must be welded at a true 90-degree angle to the other pipe connected by the elbow. *Figure 46* shows how to use the 3-4-5 method.

To fit a 90-degree angle using the 3-4-5 method, prepare the pipe and elbow as you did with previous elbows, and get the first tacks made on the 3 o'clock and 9 o'clock points. Then use the following steps:

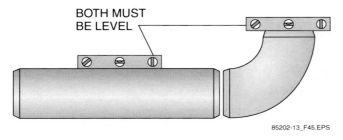

Figure 45 Aligning a 90-degree elbow using a level.

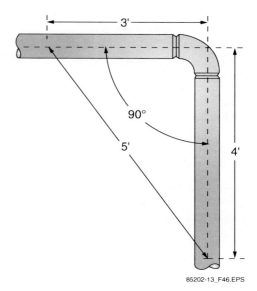

Figure 46 3-4-5 method.

Step 1 Mark the top dead center of the elbow using a center finder and the takeout formula.

Step 2 Measure 3 feet from the center of the elbow down one pipe and mark the top dead center of the pipe at this point using the center finder.

Step 3 Measure 4 feet from the center of the elbow down the other pipe, and mark the top dead center of the pipe at this point using the center finder.

Step 4 Measure the distance between these two points and adjust the pipe until this distance is 5 feet.

Step 5 Have the welder tack-weld the joint in the throat of the elbow.

> **NOTE**
> Step back and recheck alignment after each tack weld. Make sure the welder is exactly sure where to place each tack weld.

Step 6 Have the welder tack-weld the joint on the outside of the elbow.

8.5.0 Aligning a Pipe to a Flange

Flanges must be aligned to pipe both horizontally and vertically. The bolt holes must also be aligned with the other flanges on the pipe. If welding a flange to a pipe that already has a flange on one end, level the existing flange first using the two-hole method and a torpedo level.

Follow these steps to align a flange to the end of a pipe:

Step 1 Set up two adjustable pipe stands to support the pipe to be welded.

Step 2 Secure the section of pipe onto the stands with the end to be welded extending at least 12 inches beyond one of the stands.

Step 3 Level the pipe between the two stands.

Step 4 Line up the flange with the end of the pipe.

Step 5 Place a jack stand underneath the flange to help support the flange.

Step 6 Set the proper gap between the pipe and the flange.

Step 7 Place flange pins into the top two holes of the flange.

Step 8 Place a torpedo level on the flange pins.

Step 9 Adjust the flange until the two flange pins are level (*Figure 47*).

Step 10 Have the welder tack-weld the flange to the pipe at the top center of the joint.

Step 11 Remove the jack stand from underneath the flange.

Step 12 Remove the flange pins from the flange.

Step 13 Place a framing square over the flange so that the tongue of the square is flush with the face of the flange and the body of the square is aligned over the pipe (*Figure 48*).

Step 14 Adjust the flange until the distance between the body of the square and the pipe is the same the entire length of the square (*Figure 49*).

(A) FLANGE ALIGNMENT PIN

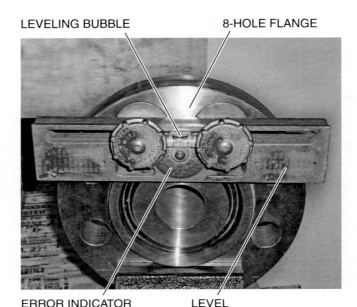

LEVELING BUBBLE   8-HOLE FLANGE

ERROR INDICATOR   LEVEL
(IN DEGREES)

(B) FLANGE WITH LEVEL

85202-13_F47.EPS

Figure 47 Aligning a flange horizontally.

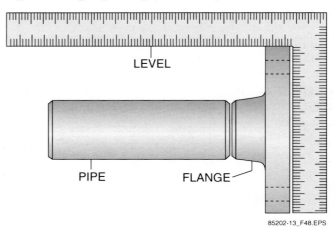

LEVEL

PIPE   FLANGE

85202-13_F48.EPS

Figure 48 Positioning a square on a flange.

Step 15 Have the welder tack-weld the sides and the bottom of the joint.

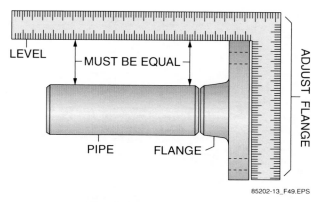

Figure 49 Adjusting flange.

The following procedure can be used when two-holing a flange to a vertical run of pipe:

Step 1 Clamp or hold a straightedge such as a framing square or pipefitter's square to two adjacent holes of the flange.

Step 2 Place a spacing tool such as a welding rod between the prepared welding surfaces.

Step 3 Establish a good reference point from which to two-hole the flange. This could be a square run of pipe, a bulkhead, or a piece of equipment.

Step 4 Check the level reading of the straight-edge and match it to the level reading of the reference point. This is a critical step.

Step 5 Place the flange with the square aligned to the two holes on the spacing tool while keeping internal pipe alignment.

Step 6 Place a level on the straightedge to match the reference point.

Step 7 Rotate the flange until the straightedge is visually aligned with the reference point.

8.6.0 Aligning a Pipe to a Tee

Two framing squares can also be used to align a tee to the end of a straight pipe. Prepare the pipe and tee as you would a 90-degree elbow and use the same procedure used to align the 90-degree ell (*Figure 50*).

Step 1 Adjust the tee until the squares are flush. When fabricating larger diameter pipe, the framing square is not large enough to use with the tee. In this case, level the pipe in the jack stands, and use a 24-inch spirit level to set the position of the tee. *Figure 51* shows how to align a tee using a level.

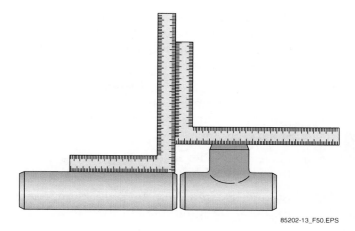

Figure 50 Pipe to tee alignment.

Step 2 Have the welder tack-weld the bottom of the joint.

Step 3 Place the square on the sides of the tee and pipe and check for misalignment from side to side.

Step 4 Have the welder tack-weld the sides of the joint.

8.7.0 Fitting Butt Weld Valves

Fitting valves in a butt-weld piping system is essentially the same process as fitting other components. For this exercise, assume that a valve is to be butt-welded to the end of a 4-inch pipe that has been beveled and cleaned and is ready to be welded. Follow these steps to fit a butt weld valve to pipe. Open the valve all the way, and remove any parts of the valve that might be damaged by the welding heat. Follow the same procedures used with the 90-degree elbow to obtain preliminary alignment with the valve stem up.

Step 1 Set the proper gap between the pipe and the valve.

Step 2 Hold the valve in line with the pipe to make a preliminary alignment (*Figure 52*).

Step 3 Have the welder make a light tack-weld at the 12 o'clock position.

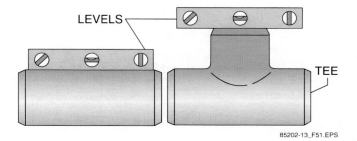

Figure 51 Aligning tee using a level.

Step 4 Place a square over the valve, with the long leg of the square aligned with the pipe and the short leg of the square flush against the valve end.

Step 5 Adjust the valve until the distance between the long leg of the square and the pipe is the same the entire length of the square.

Step 6 Have the welder make a light tack-weld at the 6 o'clock position.

Step 7 Rotate the square on the pipe 90 degrees to check the squareness of the valve from side to side.

Step 8 Adjust the valve until the distance between the long leg of the square and the pipe is the same the entire length of the square.

Step 9 Have the welder tack-weld the valve to the pipe in the 3 and 9 o'clock positions.

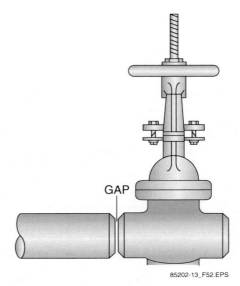

GAP

85202-13_F52.EPS

Figure 52 Preliminary valve alignment.

SUMMARY

The most important thing to remember about performing butt weld pipe fabrication is that the fit-up must be square and level. Improper fit-up can result in weld joint failure. Fit-up includes proper preparation of the pipe ends; proper alignment of the pipe ends using prescribed tools and methods; and the proper selection and use of backing rings. Alignment is accomplished using tools such as the Hi-Lo gauge, as well as levels and squares. The ability to make precise measurements is a critical skill. It is the pipefitter's job to align pipe and fittings as near perfectly as possible to produce safe systems that will give many years of service and stay maintenance-free. Pipefitters must practice the alignment procedures to perfect their skills and produce quality fit-ups every time.

1. Unless otherwise stated, elbows are always
   _____.

   a. 30 degrees
   b. 45 degrees
   c. 60 degrees
   d. 90 degrees

2. Unless the drawings specify another type,
   the type of radius elbow that should be used
   is a _____.

   a. short-radius elbow
   b. slow-radius elbow
   c. large-radius elbow
   d. long-radius elbow

3. A return bend is a _____.

   a. 0-degree fitting
   b. 45-degree fitting
   c. 120-degree fitting
   d. 180-degree fitting

4. From a straight run of pipe, a weldolet makes
   a reducing branch of _____.

   a. 30 degrees
   b. 45 degrees
   c. 60 degrees
   d. 90 degrees

5. A reducer that displaces the center line of the
   smaller of the two joining pipes is called a(n)
   _____.

   a. concentric reducer
   b. displacing reducer
   c. eccentric reducer
   d. deviating reducer

6. Flange pressure ratings vary from _____.

   a. 150 to 500 pounds
   b. 150 to 2,500 pounds
   c. 350 to 1,500 pounds
   d. 1,000 to 2,500 pounds

7. Long weld-neck flanges are normally used
   for _____.

   a. sewer and storm drain outlets
   b. vessel and tank outlets
   c. truck outlets
   d. potable water outlets

8. The flange that provides the least turbulence
   and least resistance is the _____.

   a. weld-neck flange
   b. lap-joint flange
   c. slip-on flange
   d. reducer flange

9. Lap-joint flanges are also known as _____.

   a. slip-on flanges
   b. Van Stone flanges
   c. reducing flanges
   d. raised-face flanges

10. Which of the following is a type of piping
    system drawing?

    a. Double-line
    b. Base-line
    c. Plan-line
    d. Center-line

11. A side view of a system of pipe is also called
    a(n) _____.

    a. isometric view
    b. schematic view
    c. elevation view
    d. plan view

12. A top view of a system of pipe is called a(n)
    _____.

    a. isometric view
    b. schematic view
    c. elevation view
    d. plan view

13. Detailed information about components in
    piping systems is found in the _____.

    a. isometric view
    b. title block
    c. specifications book
    d. detail drawing

14. The flat edge of a pipe that is left after a bevel
    is cut is called the _____.

    a. bevel
    b. ear
    c. face
    d. land

15. If there is a crack in a grinding wheel, the wheel _____.
    a. must be replaced
    b. can be glued
    c. can be ground down
    d. can be used at low speed

16. Stainless steel piping is cut with a(n) _____.
    a. oxyacetylene torch
    b. handsaw
    c. burn table
    d. abrasive wheel saw

17. A guillotine saw is a type of abrasive saw.
    a. True
    b. False

18. The takeout for a fitting is the distance from _____.
    a. one side to the other
    b. one face to the center of the fitting
    c. the outside to the outside
    d. one fitting to the next

19. The takeout for a long-radius elbow is always _____.
    a. twice the nominal size
    b. one-half the nominal size
    c. one and one-half the nominal size
    d. the same as the nominal size

20. The distance between a fitting on one end and the flange on the other end is determined using the _____.
    a. center-to-face method
    b. Suzuki method
    c. face-to-face method
    d. center-to-center method

21. Backing rings that provide the greatest high-low alignment are Type _____.
    a. A
    b. CCC
    c. B
    d. C

22. It is always permitted to tack-weld aligning dogs to the pipe.
    a. True
    b. False

23. Hydraulic cage clamps are used to _____.
    a. hold the grinder against the bevel
    b. realign the inner wall
    c. hold the alignment for welding
    d. support the jackstands

24. A Hi-Lo gauge is used to measure _____.
    a. the distance from the ground
    b. the diameter of the pipe
    c. misalignment of pipes
    d. the center-to-center dimension

25. When setting the gap between larger sizes of pipe, the pipe ends can be walked together by hitting the _____.
    a. other end with a ballpeen hammer
    b. jack stands in the direction you want the pipe to go
    c. pipe on top, to bounce it
    d. jack stands in the opposite direction

Trade Terms Introduced in This Module

Align: To make straight or to line up evenly.

Alloy: Two or more metals combined to make a new metal.

ASTM International: Founded in 1898, a scientific and technical organization, formerly known as the American Society for Testing and Materials, formed for the development of standards on the characteristics and performance of materials, products, systems, and services.

Bevel: An angle cut or ground on the end of a piece of solid material.

Burn-through: A hole that is formed in a weld due to improper grinding or welding.

Fit up: To put piping material in position to be welded together.

Full-penetration weld: Complete joint penetration for a joint welded from one side only.

Lateral: A fitting or branch connection that has a side outlet that is any angle other than 90 degrees to the run.

Oxide: A type of corrosion that is formed when oxygen combines with a base metal.

Root opening: The space between the pipes at the beginning of a weld; the gap is usually ⅛ of an inch.

Additional Resources

This module presents thorough resources for task training. The following resource material is suggested further study.

The Pipe Fitters Blue Book. W.V. Graves. Webster, TX: W.V. Graves Publishing Company.
The Pipe Fitter's and Pipe Welder's Handbook. Thomas W. Frankland. New York, NY: McGraw Hill.
IPT's Pipe Trades Handbook. Robert A. Lee. ITP Publishing and Training: Edmonton, Alberta, Canada.

Figure Credits

NCCER CURRICULA — USER UPDATE

NCCER makes every effort to keep its textbooks up-to-date and free of technical errors. We appreciate your help in this process. If you find an error, a typographical mistake, or an inaccuracy in NCCER's curricula, please fill out this form (or a photocopy), or complete the online form at **www.nccer.org/olf**. Be sure to include the exact module ID number, page number, a detailed description, and your recommended correction. Your input will be brought to the attention of the Authoring Team. Thank you for your assistance.

Instructors – If you have an idea for improving this textbook, or have found that additional materials were necessary to teach this module effectively, please let us know so that we may present your suggestions to the Authoring Team.

NCCER Product Development and Revision

13614 Progress Blvd., Alachua, FL 32615

Email: curriculum@nccer.org
Online: www.nccer.org/olf

❏ Trainee Guide   ❏ Lesson Plans   ❏ Exam   ❏ PowerPoints   Other _____

Craft / Level: _____   Copyright Date: _____

Module ID Number / Title: _____

Section Number(s): _____

Description: _____

_____

_____

_____

_____

Recommended Correction: _____

_____

_____

_____

_____

Your Name: _____

Address: _____

_____

Email: _____   Phone: _____

85203-13

Socket Weld Pipe Fabrication

Module Three

Trainees with successful module completions may be eligible for credentialing through NCCER's National Registry. To learn more, go to **www.nccer.org** or contact us at **1.888.622.3720**. Our website has information on the latest product releases and training, as well as online versions of our *Cornerstone* magazine and Pearson's product catalog.

Your feedback is welcome. You may email your comments to **curriculum@nccer.org**, send general comments and inquiries to **info@nccer.org**, or fill in the User Update form at the back of this module.

This information is general in nature and intended for training purposes only. Actual performance of activities described in this manual requires compliance with all applicable operating, service, maintenance, and safety procedures under the direction of qualified personnel. References in this manual to patented or proprietary devices do not constitute a recommendation of their use.

Objectives

When you have completed this module, you will be able to do the following:

1. Identify and explain socket weld fittings.
2. Read and interpret socket weld piping drawings.
3. Determine pipe lengths between socket weld fittings.
4. Describe how to fabricate socket weld fittings to pipe.

Performance Tasks

Under the supervision of the instructor, you should be able to do the following:

1. Calculate pipe lengths from line drawings using the center-to-center method, the center-to-face method, and the face-to-face method.
2. Align a socket weld elbow to the end of a pipe.
3. Align pipe to various types of fittings, to include two fittings on one pipe such as flanges and ells.

Trade Terms

Concentric reducer
Cross
Eccentric reducer
Fillet weld

Insert
Tack-weld
Straight tee
Swage

Industry-Recognized Credentials

If you are training through an NCCER-accredited sponsor, you may be eligible for credentials from NCCER's Registry. The ID number for this module is 85203-13. Note that this module may have been used in other NCCER curricula and may apply to other level completions. Contact NCCER's Registry at 888.622.3720 or go to **www.nccer.org** for more information.

Contents

Topics to be presented in this module include:

1.0.0 Introduction...1

2.0.0 Socket Weld Pipe Fittings...1

  2.1.0 Common Socket Weld Pipe Fittings.................................1

  2.2.0 Miscellaneous Socket Weld Fittings...............................1

    2.2.1 Unions...1

    2.2.2 Caps..2

    2.2.3 Reducers 3

    2.2.4 Slip-On Sleeve Joints...3

    2.2.5 Special Branch Fittings..3

  2.3.0 Socket Weld Flanges...3

3.0.0 Socket Weld Piping Drawings 3

  3.1.0 Double- and Single-Line Drawings..................................3

  3.2.0 Isometric Drawings 4

  3.3.0 Piping Symbols..4

4.0.0 Determining Pipe Lengths Between Fittings 5

  4.1.0 Center-to-Center Method 5

  4.2.0 Center-to-Face Method...5

  4.3.0 Face-to-Face Method 6

5.0.0 Fabricating Socket Weld Fittings to Pipe 6

  5.1.0 Preparing Pipe and Fittings for Alignment 7

  5.2.0 Aligning Fittings to be Welded 10

    5.2.1 Aligning 90-Degree Elbow to Pipe 10

    5.2.2 Squaring Pipe into 90-Degree Elbow.......................11

    5.2.3 Aligning Flange to Pipe..11

    5.2.4 Aligning Flange to Vertical Pipe.............................. 13

    5.2.5 Aligning 45-Degree Elbow to Pipe Using Levels 14

    5.2.6 Aligning 45-Degree Elbow Using Squares..................... 14

    5.2.7 Aligning Pipe Joined by Couplings........................... 14

  5.3.0 Installing Welded Valves.. 15

    5.3.1 Fitting Socket Weld Valves.................................... 16

  5.4.0 Installing Slip-On Welded Sleeve Joints............................. 16

Figures and Tables

Figure 1   Socket weld fittings in pipe run...1
Figure 2   Standard dimension chart 2
Figure 3   Socket weld unions 2
Figure 4   Socket weld cap..3
Figure 5   Types of reducers 4
Figure 6   Socket weld slip-on sleeve specifications 4
Figure 7   Special socket weld branch fittings..5
Figure 8   Socket weld flanges and dimensions 6
Figure 9   Two-hole method of leveling flange..7
Figure 10 Double- and single-line drawings: elevation view.............................7
Figure 11 Plan drawing, top view 7
Figure 12 Isometric drawing 8
Figure 13 Socket weld fitting symbols...9
Figure 14 Proper placement of pipe in fitting 10
Figure 15 Center-to-center piping drawing 10
Figure 16 Center-to-center method as applied to socket-weld piping 10
Figure 17 Center-to-face method 11
Figure 18 Face-to-face method...11
Figure 19 Proper placement of fitting..11
Figure 20 Location of the first tack-weld 12
Figure 21 Position of the square 12
Figure 22 Adjusting fitting...12
Figure 23 Position of the square to the pipe 12
Figure 24 Adjusting the pipe. 12
Figure 25 Plumbing a vertical pipe 13
Figure 26 Two-hole method of aligning a flange...13
Figure 27 Positioning the square on the flange...13
Figure 28 Adjusting the flange 13
Figure 29 Checking a fitting for a true 45-degree angle 14
Figure 30 Pipe to 45-degree elbow alignment..14
Figure 31 Positioning the square on the coupling...15
Figure 32 Position of squares on pipes 15
Figure 33 Proper placement of valve...16
Figure 34 Positioning a square on valves 16
Figure 35 Adjusting the valve 16
Figure 36 Position of the squares on the pipe 16
Figure 37 Positioning the square on the sleeve..17
Figure 38 Position of squares on pipes 17

Table 1   Center-to-End Dimensions and Laying Lengths..............................8

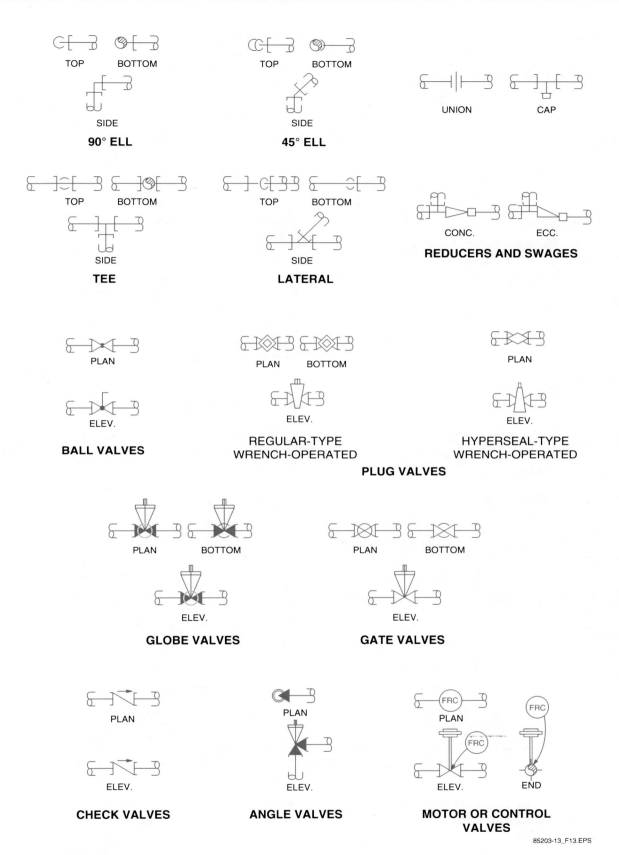

Figure 13 Socket weld fitting symbols.

85203-13_F13.EPS

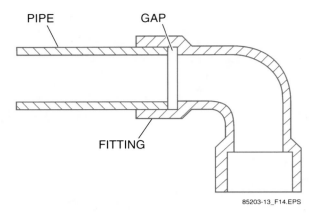

Figure 14 Proper placement of pipe in fitting.

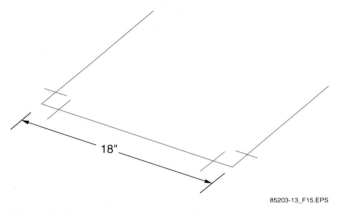

Figure 15 Center-to-center piping drawing.

Step 6 Pull the fitting away from the pipe until there is a $\frac{1}{16}$-inch gap between the edge of the socket of fitting and the scribed line. *Figure 19* shows proper placement of the fitting.

Step 7 Ensure that the gap between the line and the fitting is equal all the way around the pipe.

5.2.0 Aligning Fittings to be Welded

Before the fitting is welded to the pipe, the fittings must be properly aligned to the pipe and to the other fittings in the piping run. This can be done using squares or levels. Many of the alignment procedures require the use of both squares and levels. Squares provide more accurate alignments in the field and in other locations where the working conditions are not perfect. Perfect working conditions include being indoors with smooth, level floors and having access to accurate, quality adjustable jacks and stands. There is also a higher risk of error in using levels because levels get dropped, have weld splattered on them, and are abused due to normal wear and tear. If a level is dropped or abused, it will no longer read true. Therefore, a square gives a more accurate reading. When aligning fittings to be welded, it is critical to properly support the fabricated spool so that the fittings are not disturbed before or during weld-out. This section explains the procedures for aligning fittings to be welded using squares.

5.2.1 Aligning 90-Degree Elbow to Pipe

Follow these steps to align a 90-degree elbow to pipe. This procedure assumes there is no other fitting attached to the pipe at this time.

Step 1 Measure and cut a 2-inch carbon steel pipe 3 feet long.

Step 2 Follow the steps to prepare the pipe end and position the elbow on the pipe. The elbow should be positioned on the pipe so that the elbow turns straight up.

Step 3 Have the welder tack-weld the elbow to the pipe in the throat of the fitting at the 12 o'clock position (*Figure 20*).

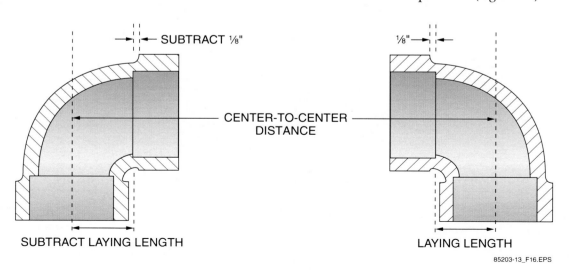

Figure 16 Center-to-center method as applied to socket-weld piping.

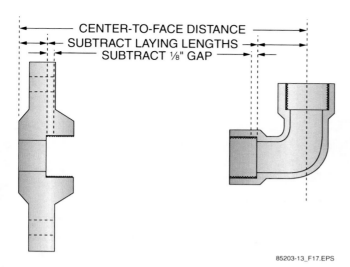

Figure 17 Center-to-face method.

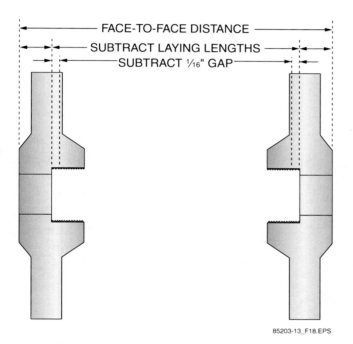

Figure 18 Face-to-face method.

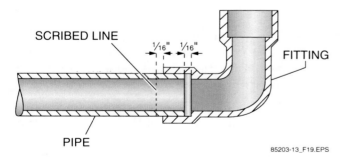

Figure 19 Proper placement of fitting.

Step 4 Place a square over the fitting so that the long leg of the square is aligned with the pipe (*Figure 21*).

Step 5 Adjust the fitting on the pipe until the distance between the pipe and the long leg of the square is the same for the entire length of the square, as shown in *Figure 22*.

> **NOTE**
>
> The elbow can also be leveled from front to back using a torpedo level, as long as the pipe is level in the pipe vise. Always remove the level before the fitting is tack-welded.

Step 6 Have the welder tack-weld the fitting to the pipe at the 6 o'clock position, align the fitting from side to side, and have the welder tack the 3 o'clock and 9 o'clock positions.

5.2.2 Squaring Pipe into 90-Degree Elbow

In this procedure, add a 3-foot length of pipe into the elbow fabricated in the previous section. To perform this procedure, the pipe and elbow should be secured in the pipe vise with the elbow turned straight up. The two pipes connected by the elbow must form a true 90-degree angle. Prepare and tack the pipe to the elbow as before. Align the pipe and fitting using the following steps:

Step 1 Hold a square against the pipe so that the long leg of the square is aligned with the pipe in the vise, as shown in *Figure 23*.

> **NOTE**
>
> Some pipefitters cut off the corner of the square so that it accommodates a weld bead.

Step 2 Adjust the pipe in the elbow until the distance between the pipe and the long leg of the square is equal for the entire length of the square (*Figure 24*).

Step 3 Have the welder tack-weld the fitting to the pipe opposite to the first tack-weld position, align the fitting from side to side, and have the welder tack the other sides.

5.2.3 Aligning Flange to Pipe

To align a flange to pipe, the flange is typically aligned horizontally using the two-hole method, and vertically, using a square. For this exercise, a flange will be fitted to the pipe that is still in the pipe vise from the previous section. Follow

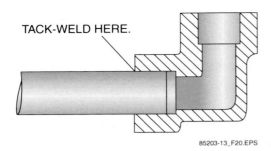

Figure 20 Location of the first tack-weld.

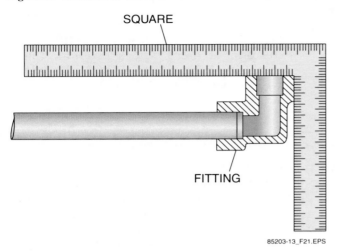

Figure 21 Position of the square.

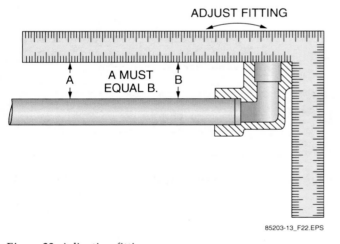

Figure 22 Adjusting fitting.

the same steps as earlier to prepare the pipe and flange and position the new flange on the pipe.

Step 1 Plumb the vertical pipe in the pipe vise from side to side (*Figure 25*).

Step 2 Use the two-hole method to align the flange to the pipe (*Figure 26*).

Step 3 Have the welder tack-weld the sides of the flange to the pipe at the 3 o'clock and the 9 o'clock positions.

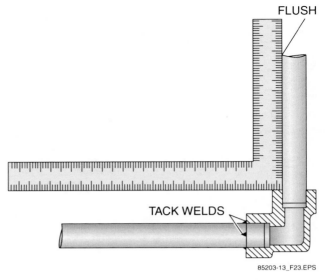

Figure 23 Position of the square to the pipe.

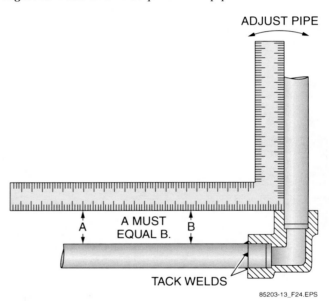

Figure 24 Adjusting the pipe.

Step 4 Remove the two-hole pins from the flange.

Step 5 Place a square over the flange so that the short leg of the square is flush with the face of the flange and the long leg of the square is aligned over the pipe (*Figure 27*).

Step 6 Adjust the flange until the distance between the long leg of the square and the pipe is the same the entire length of the square. *Figure 28* shows the flange adjustment.

Step 7 Have the welder tack-weld the top and the bottom of the pipe to the flange.

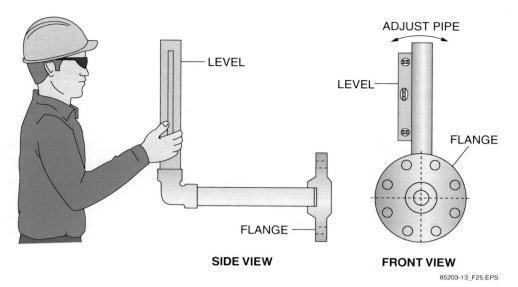

SIDE VIEW   FRONT VIEW

85203-13_F25.EPS

Figure 25 Plumbing a vertical pipe.

85203-13_F26.EPS

Figure 26 Two-hole method of aligning a flange.

5.2.4 Aligning Flange to Vertical Pipe

The two-hole method can also be used to align a flange to a vertical pipe run. The following procedure can be used when two-holing a flange to a vertical run of pipe:

Step 1 Clamp or hold a straightedge such as a framing square or pipefitter's square to two adjacent holes of the flange.

Step 2 Place a spacing tool such as a welding rod between the prepared welding surfaces.

Step 3 Establish a good reference point from which to two-hole the flange. This could

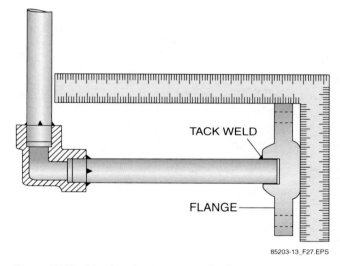

85203-13_F27.EPS

Figure 27 Positioning the square on the flange.

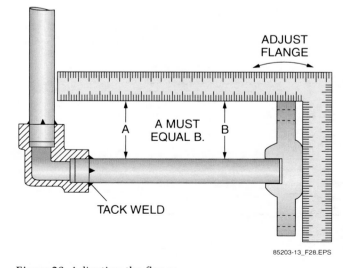

85203-13_F28.EPS

Figure 28 Adjusting the flange.

be a square run of pipe, a bulkhead, or a piece of equipment.

Step 4 Check the level reading of the straight-edge and match it to the level reading of the reference point. This is a critical step.

Step 5 Place the flange with the square aligned to the two holes on the spacing tool while keeping internal pipe alignment.

Step 6 Place a level on the straightedge to match the reference point.

Step 7 Rotate the flange until the straightedge is visually aligned with the reference point.

5.2.5 Aligning 45-Degree Elbow to Pipe Using Levels

A 45-degree elbow can be aligned to pipe, using two squares or a torpedo level with a 45-degree vial. Follow the same initial procedure as in fitting a 90-degree elbow to a pipe, and have the first tack made at the 12 o'clock position on the pipe. Then use the following steps:

Step 1 Place a torpedo level with a 45-degree vial on the face of the fitting and adjust the fitting until its face is at a true 45-degree angle (*Figure 29*).

Step 2 Remove the level from the fitting, and have the welder tack-weld the bottom of the fitting to the pipe. Check the alignment, and have the welder place the other two tacks.

5.2.6 Aligning 45-Degree Elbow Using Squares

Two squares can also be used to align a 45-degree elbow to the end of a straight pipe. Follow the same initial procedure as in fitting a 90-degree elbow to a pipe, and have the first tack made at the 12 o'clock position on the pipe. Then use the following steps to align the elbow:

Step 1 Place the short leg of the square on the top center of the straight pipe so that the long leg of the square is vertically in line with the throat of the elbow.

Step 2 Place the long leg of the second square on the face of the elbow (*Figure 30*).

Step 3 Adjust the elbow so that the same inch marks on the short and long legs of the second square touch the inner edge of the first square (see *Figure 30*).

Step 4 Have the welder tack-weld the bottom of the pipe to the elbow.

Step 5 Remove the squares from the fitting and pipe, and have the welder tack-weld the bottom of the fitting to the pipe. Check the alignment, and have the welder place the other two tacks.

5.2.7 Aligning Pipe Joined by Couplings

Straight lengths of pipe the same size can be joined by socket weld couplings. To fit a coupling, prepare the pipe and coupling as before. Have the coupling tacked to the pipe at the 12 o'clock position. Align the coupling to the first pipe using the square, as previously done with the flange, and have the welder complete the tacks (*Figure 31*).

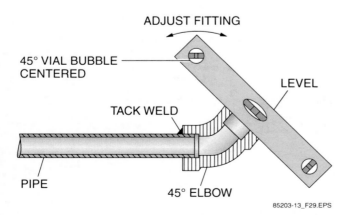

Figure 29 Checking a fitting for a true 45-degree angle.

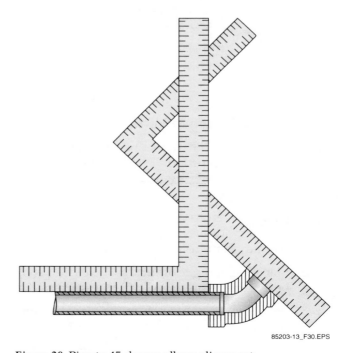

85203-13_F30.EPS

Figure 30 Pipe to 45-degree elbow alignment.

Step 1 Place the second length of pipe on jack stands in line with the first pipe and align it for the first tack-weld.

Step 2 Position a square on each pipe so that the coupling is between the long legs of each square (*Figure 32*).

Step 3 Adjust the second pipe up or down until the distance is the same between the long legs of the square at every point.

Step 4 Have the welder tack-weld the coupling to the pipe at the 6 o'clock position. Rotate the pipes 90 degrees in the jack stands, and repeat the alignment with the squares.

Step 5 Have the welder tack-weld the coupling to the pipe at the new 12 o'clock and 6 o'clock positions.

5.3.0 Installing Welded Valves

Welded valves are only available in steel and are used mainly for high-pressure, high-temperature systems that do not require frequent dismantling. The two types of welded joints used are butt weld joints and socket weld joints. Butt weld joints are available in all sizes, while socket weld joints are limited to the smaller sizes, usually less than 2 inches.

When placing the valve in the piping run in preparation for welding, ensure that the orientation of the valve is correct to allow for proper flow through the valve. Also ensure that the placement of the valve will allow sufficient clearance to operate and maintain the valve. Sometimes the piping drawings will have to be adjusted due to other equipment or piping runs in the immediate area.

Heat from welding can distort the metal of a valve. Before installing a welded valve, all components of the valve that may be damaged by the heat must be removed. These components include fiber or rubber packing and Teflon® seats. The valve must also be fully open before it is welded into place. The heat from the weld can warp the valve seats and cause the closing mechanism to mate unevenly with the seat. The welder may have to weld only a section of the joint and then

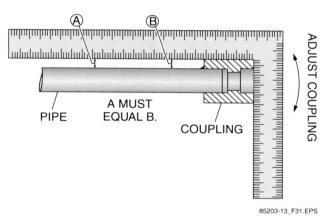

Figure 31 Positioning the square on the coupling.

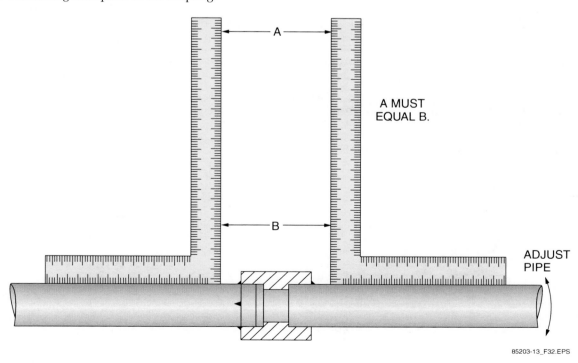

Figure 32 Position of squares on pipes.

allow the valve to cool before continuing the weld. The pipefitter's responsibilities include preparing the pipe ends and fitting the valve to the pipe.

5.3.1 Fitting Socket Weld Valves

Step 1 Prepare the pipe and valve just as was done with the coupling, and open the valve to its fully open position.

Step 2 Place the socket valve onto the end of the pipe, with the valve stem pointing straight up, ensure proper gap, and have the welder make the first tack-weld (*Figure 33*).

Step 3 Place a square over the valve, with the long leg of the square aligned with the pipe and the short leg of the square flush against the valve end (*Figure 34*).

Step 4 Adjust the valve until the distance between the long leg of the square and the pipe is the same the entire length of the square (*Figure 35*).

Step 5 Have the welder tack-weld the valve to the pipe in the 6 o'clock position.

Step 6 Rotate the framing square on the valve so that the long leg of the square is at the center line of the side of the pipe.

Step 7 Repeat Step 4 with the square in this position.

Step 8 Have the welder tack-weld the valve to the pipe at the 3 o'clock and 9 o'clock positions.

Step 9 Prepare, insert and get a first tack-weld on the other pipe in the open side of the valve, and, as it was done with the coupling, align the valve and pipes with two squares (*Figure 36*).

Step 10 Have the welder make the other tacks.

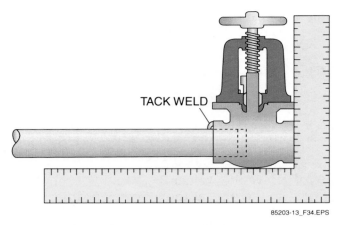

Figure 34 Positioning a square on valves.

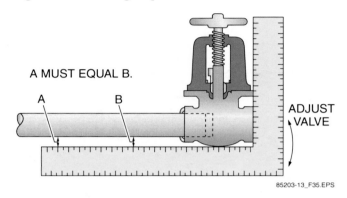

Figure 35 Adjusting the valve.

5.4.0 Installing Slip-On Welded Sleeve Joints

To join two lengths of pipe using a slip-on sleeve, the sleeve must be aligned horizontally and vertically to each of the joining pipes. For this exercise, two pipes will be joined using a slip-on sleeve.

Step 1 Follow the steps to prepare the pipe end and place the pipe in a vise. Ensure that the pipe is level horizontally before proceeding.

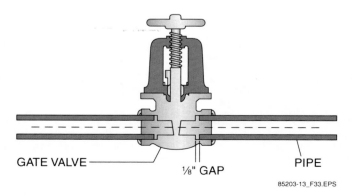

Figure 33 Proper placement of valve.

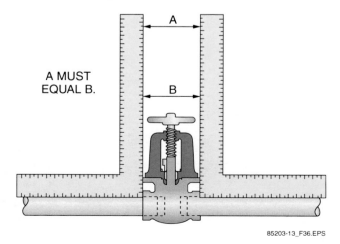

Figure 36 Position of the squares on the pipe.

NCCER – *Maritime Pipefitting Level Two* 85203-13

Step 2 From the end of the pipe, measure one-half the length of the sleeve and subtract ¹⁄₁₆". Scribe a mark on the pipe at this point.

Step 3 Place the sleeve over the end of the pipe, aligning it with the scribed mark. Have the welder tack-weld the sleeve to the pipe at the 12 o'clock position.

Step 4 Align the sleeve to the first pipe using the square, as was done with the coupling in the previous section, and have the welder complete the tacks (*Figure 37*).

Step 5 Place the second length of pipe on jack stands in line with the sleeve tack-welded to the first pipe and repeat the steps used for the first pipe.

Step 6 Position a square on each pipe so that the sleeve is between the long legs of each square (*Figure 38*).

Step 7 Adjust the second pipe up or down until the distance is the same between the long legs of the square at every point.

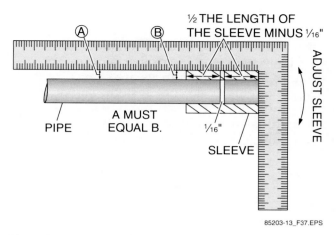

Figure 37 Positioning the square on the sleeve.

Step 8 Have the welder tack-weld the coupling to the pipe at the 6 o'clock position. Rotate the pipes 90 degrees in the jack stands, and repeat the alignment with the squares.

Step 9 Have the welder tack-weld the coupling to the pipe at the new 12 o'clock and 6 o'clock positions.

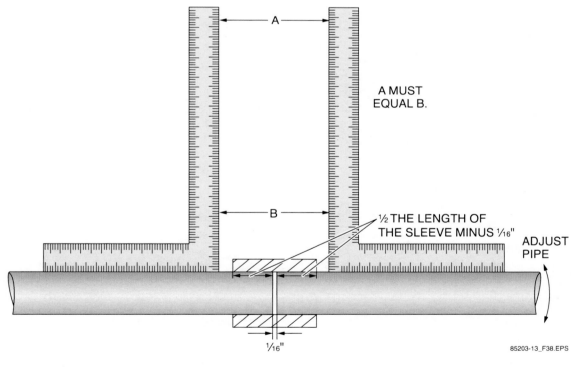

Figure 38 Position of squares on pipes.

SUMMARY

Socket welding is a very common method used to join pipe in the field. The proper pipe end preparation and fit-up procedures are critical to producing a quality, leakproof system. The purpose of this module is to teach methods of aligning socket fittings to pipe that will be welded to produce quality joints. There are other methods to perform many of these alignment procedures that will be encountered in the field. The most important thing to remember is that regardless of the procedure used to align the fitting, the end result must be a fit-up that is square and level. Through practice, pipefitters will be able to rapidly and properly fabricate socket weld piping systems.

1. In socket weld pipe systems, the pipe is joined by inserting it into a socket or well on the fitting and _____.
   a. tack-welding the pipe to the fitting
   b. butt-welding the pipe to the fitting
   c. fillet-welding the pipe to the fitting
   d. spot-welding the pipe to the fitting

2. The maximum pipe size used in socket weld applications is _____.
   a. 3 inches
   b. 4½ inches
   c. 6 inches
   d. 10 inches

3. Pressure classes of pipe for socket weld fittings include _____.
   a. 2,000, 3,000, 6,000, and 9,000 pounds
   b. 3,000, 4,000, 5,000, and 6,000 pounds
   c. 4,000, 6,000, 7,000, and 9,000 pounds
   d. 6,000, 7,000, 8,000, and 9,000 pounds

4. Which of the following is one of the most common socket weld fittings?
   a. Sockolet
   b. Elbolet
   c. Latrolet
   d. Cross

5. In socket weld applications, a swage is a _____.
   a. socketless reducer
   b. flaring tool
   c. special elbow
   d. union

6. The welded coupling that makes a 45-degree reducing branch from a straight run of pipe is a(n) _____.
   a. reducing tee
   b. latrolet
   c. sockolet
   d. elbolet

7. Which of the following is a class of flange?
   a. 250 pound
   b. 400 pound
   c. 1,500 pound
   d. 4,000 pound

8. The purpose of two-holing a flange is to _____.
   a. reduce the number of holes
   b. align the bolt holes
   c. increase the number of holes
   d. set the face at perpendicular to the pipe

9. To find the maximum pressure for a part of the system shown on a drawing, look in the _____.
   a. title block
   b. specifications
   c. pipe markings
   d. schematics

10. The length of the pipe between socket weld fittings can be calculated by finding the design length from center to center and subtracting the laying length and the _____.
    a. required gap
    b. takeout
    c. pipe diameter
    d. number of fittings

11. The laying length of a socket weld fitting is the length _____.
    a. from the fitting face to the end of the pipe inside the socket
    b. of the fillet
    c. from the face of the socket to the very back of the socket
    d. from the center of the fitting to the back of the socket

12. A Gap-A-Let® is used to set a standard gap length from _____.
    a. the back of the socket to the face of the socket
    b. the back of the socket to the end of the pipe inside
    c. the outside of the pipe to the inside of the socket
    d. one pass of the fillet to the next pass

13. The center-to-center method is used to determine the correct length of cut pipe between two _____.

    a. valves
    b. flanges
    c. fittings
    d. pipe caps

14. The face-to-center method is used to determine the correct length of cut pipe between

    _____.

    a. valves and fittings
    b. caps and flanges
    c. two fittings
    d. a fitting and a flange

15. The face-to-face method is used to determine the correct length of cut pipe between _____.

    a. valves and fittings
    b. two fittings
    c. two flanges
    d. a fitting and a flange

Trade Terms Introduced in This Module

Concentric reducer: A reducer that maintains the same center line between the two pipes that it joins.

Cross: A fitting with four branches all at right angles to each other.

Eccentric reducer: A reducer that displaces the center line of the smaller of the two joining pipes to one side.

Fillet weld: A weld with a triangular cross section joining two surfaces at right angles to each other.

Insert: A type of reducer that fits into the socket of a fitting to reduce the line size.

Tack-weld: A weld made to hold parts together in proper alignment until the final weld is made.

Straight tee: A fitting that has one side outlet 90 degrees to the run.

Swage: A type of socketless fitting in which one side is larger than the other.

Additional Resources

This module presents thorough resources for task training. The following resource material is suggested further study.

The Pipe Fitters Blue Book. W.V. Graves. Webster, TX: W.V. Graves Publishing Company.

Figure Credits

Courtesy of Mathey Dearman, Figure 26

NCCER CURRICULA — USER UPDATE

NCCER makes every effort to keep its textbooks up-to-date and free of technical errors. We appreciate your help in this process. If you find an error, a typographical mistake, or an inaccuracy in NCCER's curricula, please fill out this form (or a photocopy), or complete the online form at **www.nccer.org/olf**. Be sure to include the exact module ID number, page number, a detailed description, and your recommended correction. Your input will be brought to the attention of the Authoring Team. Thank you for your assistance.

Instructors – If you have an idea for improving this textbook, or have found that additional materials were necessary to teach this module effectively, please let us know so that we may present your suggestions to the Authoring Team.

NCCER Product Development and Revision

13614 Progress Blvd., Alachua, FL 32615

Email: curriculum@nccer.org
Online: www.nccer.org/olf

❑ Trainee Guide   ❑ Lesson Plans   ❑ Exam   ❑ PowerPoints   Other _____

Craft / Level: _____   Copyright Date: _____

Module ID Number / Title: _____

Section Number(s): _____

Description: _____

_____

_____

_____

_____

Recommended Correction: _____

_____

_____

_____

_____

Your Name: _____

Address: _____

_____

Email: _____   Phone: _____

85204-13

Brazing

Module Four

Trainees with successful module completions may be eligible for credentialing through NCCER's National Registry. To learn more, go to **www.nccer.org** or contact us at **1.888.622.3720**. Our website has information on the latest product releases and training, as well as online versions of our *Cornerstone* magazine and Pearson's product catalog.

Your feedback is welcome. You may email your comments to **curriculum@nccer.org**, send general comments and inquiries to **info@nccer.org**, or fill in the User Update form at the back of this module.

This information is general in nature and intended for training purposes only. Actual performance of activities described in this manual requires compliance with all applicable operating, service, maintenance, and safety procedures under the direction of qualified personnel. References in this manual to patented or proprietary devices do not constitute a recommendation of their use.

Objectives

When you have completed this module, you will be able to do the following:

1. Identify the tools and materials used to braze copper and other metals.
2. Describe how to prepare pipe ends for brazing.
3. Describe the materials and methods used to braze piping made of copper and other metals including stainless steel, brass, cupronickel, nickel-copper, valve bronze, and Inconel.

Performance Task

Under the supervision of the instructor, you should be able to do the following:

1. Properly prepare copper or cupronickel pipe and fittings for brazing.

Trade Terms

Acetone
Alloy
Brazing ring
Brazing
Capillary action
Cup depth

Fillet
Flashback arrestor
Fluxes
Nonferrous
Oxidation
Wetting

Industry-Recognized Credentials

If you are training through an NCCER-accredited sponsor, you may be eligible for credentials from NCCER's Registry. The ID number for this module is 85204-13. Note that this module may have been used in other NCCER curricula and may apply to other level completions. Contact NCCER's Registry at 888.622.3720 or go to **www.nccer.org** for more information.

Contents

Topics to be presented in this module include:

1.0.0 Introduction...1
1.1.0 Brazing Safety 1
2.0.0 Brazing Equipment and Materials..2
2.1.0 Pipe Cutting and Cleaning Equipment 7
2.2.0 Materials...8
2.2.1 Filler Metals 8
2.2.2 Fluxes 9
3.0.0 Preparing Pipe Ends for Brazing 10
4.0.0 Brazing Pipe Fittings 12
4.1.0 Brazing Equipment Setup...12
4.2.0 Basic Procedure to Prepare Oxyacetylene Brazing
        Equipment for Use 13
4.3.0 Lighting the Oxyacetylene Torch...15
4.3.1 Types of Flames...16
4.3.2 Lighting Air-Acetylene Equipment 18
4.4.0 Brazing Joints 19
4.4.1 Brazing Metals Other Than Copper 21
4.4.2 Brazing Dissimilar Metals 21

Figures and Tables

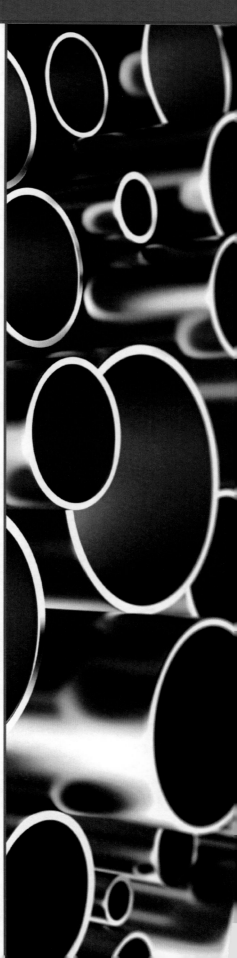

Figure 1 Capillary action...1
Figure 2 Acetylene cylinder with a valve safety cap 2
Figure 3 An oxyacetylene brazing setup 3
Figure 4 An air-acetylene brazing setup 4
Figure 5 Portable oxyacetylene equipment....................................4
Figure 6 Oxygen and fuel gas regulators 5
Figure 7 Torch wrench..6
Figure 8 Medium- and light-duty brazing torch handles.................6
Figure 9 Flashback arrestor 6
Figure 10 Brazing tips 6
Figure 11 Torch tip cleaners 7
Figure 12 Cup-type striker 7
Figure 13 Tubing cutters 7
Figure 14 Tubing cleaning brush 8
Figure 15 Pipe reamer..8
Figure 16 BCuP-5 brazing rods 9
Figure 17 Example of a brazing ring 9
Figure 18 A high-performance brazing flux 9
Figure 19 Measurements needed to properly size and cut tubing...............10
Figure 20 Preparation for brazing 11
Figure 21 Typical empty cylinder marking 12
Figure 22 Clamshell and ring caps 13–14
Figure 23 Inspecting the fittings...14
Figure 24 Installing the regulator with the special torch wrench 14
Figure 25 Types of flames...17
Figure 26 Reduce the acetylene flow until the flame returns to the tip........18
Figure 27 Transducer, top and bottom safety plugs.......................19
Figure 28 Typical welding goggles 19
Figure 29 Multi-flame, or rosebud, heating tip...............................20
Figure 30 Working in overlapping sectors 20

Table 1 Tip Sizes Used for Common Tubing Sizes........................7
Table 2 Brazing Filler Metals...8
Table 3 Silver-Brazed Fitting Takeout Dimensions (400# WOG) 10
Table 4 Brazing Tip Operational Data..16

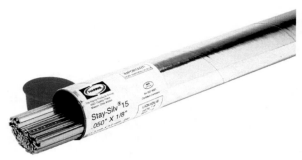

Figure 16 BCuP-5 brazing rods.

- *BCuP-5* – Brazing copper or cupronickel to stainless steel. As a general rule, stainless steel is represented in the joint when BCuP-5 is used.
- *BCuP-4* – Brazing materials or combinations of materials not listed above.

Most brazing situations in maritime pipefitting call for using a brazing ring (*Figure 17*). However, brazed joints on pipe sizes of ³⁄₈" and smaller do not typically use brazing rings. Brazing rings are made of filler metal formed in a circle and are of the same composition as other filler metals discussed here. The rings are designed to be installed inside of fittings. Brazing rings allow a consistent amount of filler to be applied to each joint.

Brazing rings are available in sizes to match common copper tube sizes from ¼ to 1¹⁄₈ inch (6.4 mm to 28.6 mm). A split in the ring allows it to be inserted tightly into the fitting. Fluxes are always used with brazing rings in maritime applications unless otherwise specified.

2.2.2 Fluxes

Fluxes are required for brazing work. A flux performs the following functions:

- It chemically cleans and protects the surfaces of the tubing and fitting from oxidation. Oxidation is a chemical process that occurs when the oxygen in the air combines with the recently cleaned metal. Oxidation produces tarnish or rust in metal and prevents the filler metal from adhering. The oxidizing process is accelerated when the tube and fitting are heated.
- It allows the filler metal to flow more easily into the joint by promoting wetting of the metals. Wetting is the spreading of the filler material thoroughly over the base metal.
- It floats out remaining oxides ahead of the molten filler metal.

Brazing fluxes like the one shown in *Figure 18* are different from soldering fluxes because of the higher temperatures used. Brazing fluxes are also more corrosive than soldering fluxes so care must be taken never to use one for the other. For best results, use the flux recommended by the manufacturer of the brazing filler metals. Since brazing temperatures are significantly higher than those for soldering, oxides form even more rapidly. Without the proper use of flux when needed, the brazed joint will not reach an acceptable level of quality or strength. Brazed joints often must be disassembled on shipboard. Flux should be applied to these joints as well, before they are heated.

Fluxes are required for brazing. Alloy filler metals that contain phosphorus, such as BCuP-3 through BCuP-5, are considered self-fluxing when used in clean copper-to-copper applications. However, in the maritime environment, a flux is still applied to copper-to-copper joints, regardless of the filler metal composition. A flux is also used with brass, bronze, and ferrous metals. Flux application is discussed later in this module.

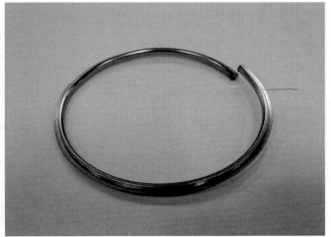

Figure 17 Example of a brazing ring.

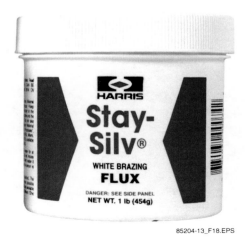

Figure 18 A high-performance brazing flux.

3.0.0 PREPARING PIPE ENDS FOR BRAZING

Preparation is the key to producing a solid, leakproof brazed joint. Note that the procedure here assumes a face-to-face measurement is being made. For the center-to-center method, the takeout must be subtracted from the measurement. *Table 3* provides the takeout dimensions and cup depths for 400# WOG fittings, which are the most commonly brazed. Use the following procedure to prepare the tubing and fittings for brazing.

Step 1 Measure the distance between the faces of the two fittings.

Step 2 Determine the **cup depth** (depth of insertion of the tube into the fitting) of each of the fittings. If the fitting has an internal shoulder, measure from the shoulder.

Step 3 Add the cup depth of both fittings to the measurement found in to find the length of tubing needed (*Figure 19*).

Step 4 Cut the copper tubing to the correct length. Cuts must be square (90-angle) between the tubing and the cut. In maritime applications, the tubing is almost always cut with a saw to avoid forming an internal ridge.

NOTE

To obtain proper clearances for capillary action, the tubing may have to be expanded. This is especially true when a tubing cutter is used, as it tends to collapse the tubing slightly as the blade penetrates the tube wall. Pipe expansion tools are readily available. Most of these tools consist of a set of expandable collars. The correct size collar is inserted in the pipe, and then the collar expansion tool handle is turned until turning becomes difficult. Some toolmakers recommend that the pipe be heated slightly before using the tool.

Step 5 Ream the inside and outside of both ends of the copper tubing using the proper size reamer. There must be no ridge on the end of the tubing, inside or out.

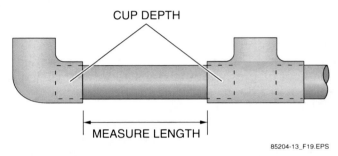

Figure 19 Measurements needed to properly size and cut tubing.

Table 3 Silver-Brazed Fitting Takeout Dimensions (400# WOG)

SIZE	SR 90	LR 90	45 ELB	CPLG	CUP DEPTH
1/4"	7/16"	1 5/16"	1/4"	3/16"	1/4"
3/8"	1/2"	1 3/16"	5/16"	3/16"	5/16"
1/2"	5/8"	1 3/4"	3/8"	1/4"	3/8"
3/4"	3/4"	1 9/16"	7/16"	5/16"	3/8"
1"	1"	1 7/8"	5/8"	3/8"	7/16"
1 1/4"	1 1/8"	2 1/4"	3/4"	7/16"	1/2"
1 1/2"	1 3/16"	2 1/2"	3/4"	5/16"	5/8"
2"	1 7/16"	3 1/16"	3/4"	7/16"	5/8"
2 1/2"	1-7/8"	3 11/16"	1 1/8"	5/8"	3/4"
3"	2 1/4"	4 9/16"	1 5/16"	3/4"	1 3/16"
3 1/2"	2 9/16"	5 5/16"	1 1/2"	13/16"	7/8"
4"	2 7/8"	6 1/32"	1 11/16"	1 5/16"	7/8"
5"	3 1/2"	7 1/8"	2 1/16"	1 1/16"	1"
6"	4"	7 7/8"	2 5/16"	1 1/4"	1 1/16"
8"	5 1/4"	8 11/16"	2 15/16"	1 9/16"	1 5/16"
10"	6 9/16"	9 1/2"	2 5/8"	1 3/4"	1 1/2"
12"	7 7/8"	N/A	4 15/16"	1 7/8"	1 5/8"

Take care when cleaning the tubing and fittings to remove all the abrasions on the copper without removing a large amount of metal. Abrasions can weaken or ruin a copper joint. Do not touch or brush away filings from the tube or fitting with your fingers because your fingers will also contaminate the freshly cleaned metal. Also, filings and burrs can also easily injure your fingers. The tubing and fitting should be thoroughly cleaned with alcohol or a suitable degreaser.

Step 6 Clean the inside of the fitting cup(s) with a brush as shown in *Figure 20A*.

Step 7 Clean the outside of the tubing to a bright finish using No. 00 steel wool, a piece of emery cloth, or an abrasive pad. Refer to *Figure 20B* and *Figure 20C*. The tube should be cleaned back from the end slightly farther than the cup depth of the fitting.

Step 8 Examine the fitting and tube to ensure they are well cleaned and no significant flaws exist in the finish. Flaws in the tubing can usually be seen more easily after cleaning, since oxidation may remain in a dent or gouge.

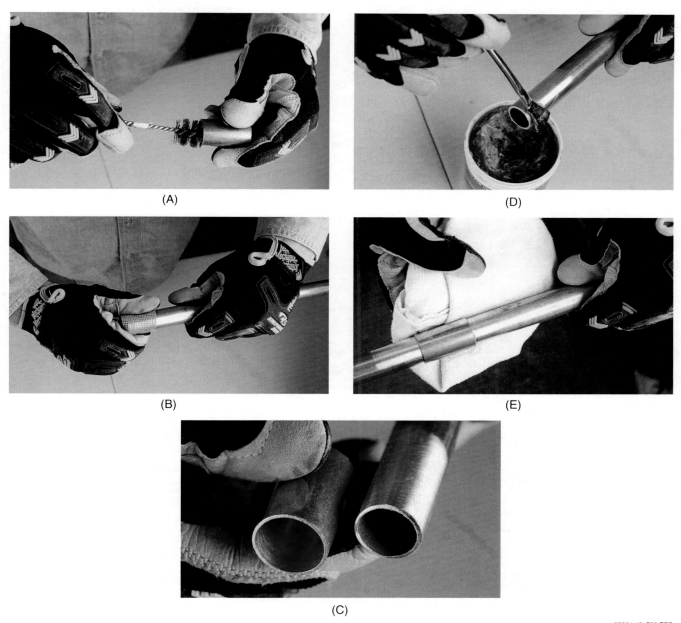

(A)

(D)

(B)

(E)

(C)

85204-13_F20.EPS

Figure 20 Preparation for brazing.

Step 9 Scribe or lightly punch four evenly spaced marks around the tubing to use as an indication that the tubing is fully seated in the fitting. The marks should be made at the cup depth plus 1" from the end of the tubing.

Step 10 Apply flux (if required) to the end of the copper tubing before inserting it into the fitting (*Figure 20D*). Apply flux to the interior of the fitting cup.

Step 11 Insert the tube into the fitting socket, and push and turn the tube into the socket until the tube touches the inside shoulder of the fitting. If the fitting does not have a shoulder, make sure that the tubing has been inserted far enough to make a good joint.

Step 12 Wipe away any excess flux from the outside of the joint (*Figure 20E*).

Step 13 Recheck the tube and fitting for proper alignment. Measure the distance from the face of the fitting to the marks made in Step 9; the distance should be 1". If necessary, support the fitting so it does not fall out of alignment during brazing. Also make sure that the joint is positioned so there is no pull-back (backing out of the tube from inside of the fitting) during the brazing process. The tube must be fully seated in the fitting.

4.0.0 BRAZING PIPE FITTINGS

Once the tubing and fitting have been prepared, the joint is ready for brazing. The following sections outline the brazing process.

4.1.0 Brazing Equipment Setup

Most brazed joints are made at temperatures between 1,200°F and 1,550°F (649°C to 843°C). This is the range of temperatures where the popular brazing alloys are fluid enough to flow, without becoming too fluid. The torch flame temperature must be significantly higher than this to achieve good control over the heating process. Because of the higher temperatures needed, oxyacetylene equipment is typically used for brazing and is the primary approach covered here.

During use, transportation, and storage, oxygen and acetylene cylinders must be secured with a stout cable or chain in the upright position to prevent them from falling and injuring

85204-13_F21.EPS

Figure 21 Typical empty cylinder marking.

people or damaging equipment. When stored at the job site, oxygen and acetylene cylinders must be stored separately with at least 20 feet (6 m) between them, or with a 5 feet (1.5 m) high, 30-minute minimum firewall separating them. As a general rule, cylinders should be stored without a regulator attached, and the regulator should be removed when a cylinder is moved. With a regulator attached, the chances of shearing off the cylinder valve stem are increased, since it protrudes well beyond the body of the cylinder. Store empty cylinders away from partially full or full cylinders and make sure they are properly marked to clearly show that they are empty (*Figure 21*). Mark the cylinder with "MT" and the date. Before leaving the area make sure that unused cylinders are securely turned off. If work is being done onboard or in a block, remove the tanks and hoses from the workspace.

High-pressure cylinders can also be equipped with a clamshell cap (*Figure 22A*) which can be closed to protect the cylinder valve with or without a regulator installed on the valve. This enables safe movement of the cylinder after the cylinder valve is closed. This type of cap is usually secured to the cylinder body cap threads when it is installed so that it cannot be removed. When the clamshell is closed, it can also be padlocked to prevent unauthorized operation of the cylinder valve.

Acetylene cylinders can be equipped with a ring guard cap, such as the one in *Figure 22B*, that protects the cylinder valve with or without a regulator installed on the valve. This enables safe movement of the cylinder after the cylinder valve is closed. This type of cap is usually secured to the cylinder body cap threads when it is installed so that it cannot be removed.

4.2.0 Basic Procedure to Prepare Oxyacetylene Brazing Equipment for Use

Step 1 Install and securely fasten the oxygen and acetylene cylinders in a bottle cart or in an upright position against a wall or other substantial structure. Acetylene cylinders must be used in the upright position. Otherwise, liquid acetone could be pulled out of the tank.

Step 2 Install the oxygen regulator on the oxygen cylinder.
- Remove the cylinder protective cap.
- Open (crack) the oxygen cylinder valve just long enough to allow a small amount of oxygen to pass through and blow out any debris, then close it.
- Carefully inspect the regulator attaching threads, cylinder threads, and mating surfaces for damage and cleanliness (*Figure 23*).
- Turn the adjusting screw on the oxygen regulator counterclockwise (out) until it is loose. This will shut off the regulator output and prevent accidental overpressurizing of the hose and torch during hookup.
- Using a torch wrench, install the oxygen regulator on the cylinder (*Figure 24*). Oxygen cylinders and regulators have right-hand threads. Tighten the

nut snugly. Be careful not to over-tighten the nut because this may strip the threads.

CLAMSHELL OPEN TO ALLOW
CYLINDER VALVE OPERATION

LATCH PIN
(OR PADLOCK)

CLAMSHELL CLOSED FOR MOVEMENT OR
PADLOCKED TO PREVENT OPERATION
OF CYLINDER VALVE

CLAMSHELL CLOSED FOR TRANSPORT

85204-13_F22A.EPS

Figure 22 Clamshell and ring caps. (1 of 2)

85204-13_F22B.EPS

Figure 22 Clamshell and ring caps. (2 of 2)

Step 3 Install the acetylene regulator on the acetylene cylinder.
- Remove the cylinder protective cap.
- Open (crack) the acetylene cylinder valve, using the valve wrench, just long enough to allow a small amount of acetylene to pass through the valve, then close it.
- Turn the adjusting screw on the acetylene regulator counterclockwise until it is loose. This will shut off the regulator output and prevent accidental overpressurizing of the hose and torch during hookup.
- Using a torch wrench, install the acetylene regulator on the cylinder. Acetylene cylinders and regulators have left-hand threads. Tighten the nut snugly. Be careful not to overtighten the nut because this may strip the threads.

Step 4 Connect the hoses and brazing torch.
- If they are not already in place or a part of the torch handle, install flashback arrestors on the oxygen and acetylene hoses at the torch handle inlets.

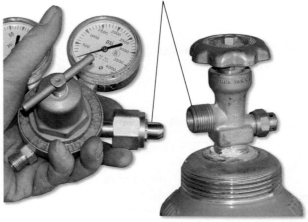

85204-13_F23.EPS

Figure 23 Inspecting the fittings.

- Connect the green hose to the oxygen regulator and the red hose to the acetylene regulator. Tighten the hoses snugly. Be careful not to over tighten the fittings because this may strip the threads.

WARNING!

Do not allow anyone to stand to the front or rear of the oxygen regulator when opening the cylinder valve, in case the regulator fails suddenly. Open the oxygen cylinder valve slowly.

Step 5 Purge (clean) the oxygen hose.
- Crack the oxygen cylinder valve slowly until the pressure stops rising on the cylinder, or high-pressure, gauge. This now indicates the pressure in the cylinder. Then slowly open the valve completely.

TORCH WRENCH

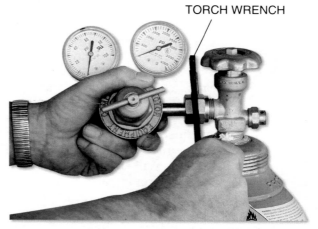

85204-13_F24.EPS

Figure 24 Installing the regulator with the special torch wrench.

- Turn the oxygen regulator adjusting screw clockwise (in) until a small amount of pressure (3 to 5 psig or 144 to 239 kPa) shows on the oxygen working-pressure gauge. Allow a small amount of pressure to build up and purge the oxygen hose, clearing it of any loose debris.
- Turn the oxygen regulator adjusting screw counterclockwise (out) until it is again loose. This will shut off the regulator output.

Step 6 Purge (clean) the acetylene hose.
- Crack the acetylene cylinder valve open slowly until the cylinder pressure registers on the acetylene cylinder, or high-pressure, gauge. Then open it about ½ to ¾ of a turn.
- Turn the acetylene regulator adjusting screw clockwise (in) until a small amount of pressure shows on the acetylene working-pressure gauge. Allow a small amount of pressure to build up and purge the acetylene hose, clearing it of any loose debris.
- Turn the acetylene regulator adjusting screw counterclockwise (out) until it is again loose. This will shut off the regulator output.

Step 7 If not already in place, install a flashback arrestor on the acetylene inlet of the torch.

Step 8 Install the brazing torch on the ends of the hoses and close the valves on the torch. Remember that the oxygen hose has normal right-hand threads while the acetylene hose has left-hand threads.

Step 9 Check the oxyacetylene equipment for leaks.
- Adjust the acetylene regulator adjusting screw for 10 psig (479 kPa) on the working-pressure gauge.
- Adjust the oxygen regulator adjusting screw for 20 psig (958 kPa) on the working-pressure gauge.

- Close the oxygen and acetylene cylinder valves and check for leaks while the pressure remains in the hoses and torch handle. If the working-pressure gauges remain at 10 and 20 psig (479 and 958 kPa) for several minutes, there should be no significant leaks in the system. If the readings drop, there is a leak. Use a soap solution or commercial leak detection fluid to check the oxygen or acetylene connections for leaks. Note that the leak could also be on the regulator assembly. If the regulator assembly is found to be leaking, stop using it immediately. Do not attempt to repair it in the field.
- Open both valves on the torch handle to release the pressure in the hoses. Watch the working-pressure gauges until they register zero, then close the valves on the torch handle.
- Turn the oxygen and acetylene regulator valves counterclockwise until they are loose. This will release the spring pressure on the regulator diaphragms and completely close the regulators.

4.3.0 Lighting the Oxyacetylene Torch

After the oxyacetylene brazing equipment has been properly prepared, the torch can be lit and the flame adjusted for brazing. Before beginning however, it is important to know the correct oxygen and acetylene operating pressures for the tip in use. *Table 4* provides one example of a manufacturer's table containing this data for a particular series of tips. It is important that the information for the actual tip in use be consulted. This table should not be considered a general guide for all tips. For hose lengths greater than 25 feet (7.6 m), the pressure at the regulator is generally increased 2 to 3 psig (96 to 144 kPa) to compensate for pressure drop. Remember, however, that the acetylene pressure should never exceed 15 psig (718 kPa) while the gas is flowing.

Table 4 Brazing Tip Operational Data

Part #	UPC #	Acetylene Consumption SCFH	Fuel Pressure psig	Oxygen Pressure psig	Welding Metal Thickness	Brazing Copper Tubing
Type 17-000	30250	1–2	5	5	1/32"	1/8"
Type 17-00	30251	2–3	5	5	3/64"	1/4"
Type 17-0	30252	2–4	5	5	5/64"	1/2"
Type 17-1	30253	3–6	5	5	3/32"	3/4"
Type 17-2	30254	5–10	5	5	1/8"	1"
Type 17-3	30255	8–18	6	7	3/16"	1½"
Type 17-4	30256	10–25	7	10	1/4"	2"
Type 17-5	30257	15–35	8	12	1/4"–1/2"	3"
Type 17-6	30258	25–45	9	14	1/2"–3/4"	4"
Type 17-7	30259	30–60	10	16	3/4"–1¼"	6"

> **WARNING!**
> Post a fire watch while brazing in case material near the work ignites. When working close to flammable material such as wood, use a fire blanket or other flame/heat blocking material to protect it. Always braze in a well-ventilated area because fumes from the process can irritate your eyes, nose, throat, and lungs. Wear respiratory equipment if necessary.

> **WARNING!**
> When preparing to braze, make sure you are wearing clothing made of non-flammable fabric such as cotton. Avoid synthetic fabrics such as nylon, which can melt and adhere to the skin. Wear a long-sleeve shirt to reduce the chance of burns from sparks or molten metal. Pants should not have cuffs. Wear shaded eye protection per OSHA requirements, as well as face protection, flame-resistant gloves and high-top work boots.

4.3.1 Types of Flames

There are three types of flames: neutral, carburizing (reducing), and oxidizing. *Figure 25* shows the differences in the types of flames. The neutral flame burns equal amounts of oxygen and acetylene. The inner cone is bright blue in color, surrounded by a fainter blue outer flame envelope that results when the oxygen in the air combines with the superheated gases from the inner cone. A neutral flame is used for almost all fusion welding or heavy brazing applications.

A carburizing, or reducing, flame has a white feather created by excess fuel. The length of the feather depends on the amount of excess fuel in the flame. The outer flame envelope is brighter than that of a neutral flame and is much lighter in color. The excess fuel in the carburizing flame produces large amounts of carbon. The carburizing flame is cooler than the neutral flame and is used for light brazing to prevent melting of the base metal.

An oxidizing flame has an excess amount of oxygen. Its inner cone is shorter, with a bright blue edge and a lighter center. The cone is also more pointed than the cone of a neutral flame. The outer flame envelope is very short and often fans out at the ends. The hottest flame, it is sometimes used for brazing cast iron or other metals.

Use the following procedure to light an oxyacetylene torch:

Step 1 Set up the oxyacetylene torch as discussed previously, heeding all warnings. Make sure the desired tip is installed before lighting the torch. Be aware of the needed gas pressure settings for the tip in use.

Step 2 Adjust the torch oxygen flow.
- Open the oxygen cylinder valve slightly until pressure registers on the oxygen cylinder high-pressure gauge and stops rising. At that point, open the valve fully.
- Turn the oxygen regulator adjusting screw clockwise until pressure shows on the oxygen working-pressure gauge. Adjust the regulator for the manufacturer's recommended pressure setting.
- Open the oxygen valve on the torch handle.
- Check the regulator and ensure that the pressure remains at the desired setting. Adjust as needed.
- Close the oxygen valve on the torch handle.

CARBURIZING FLAME

NEUTRAL FLAME

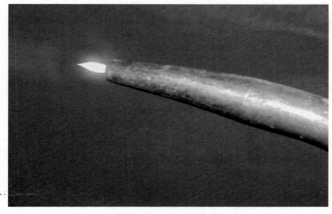
OXIDIZING FLAME

85204-13_F25.EPS

Figure 25 Types of flames.

> **NOTE**
>
> Always fine-tune the pressure setting with the torch valve open. When it is closed, the pressure may register slightly higher. Regulators are incapable of making pressure adjustments without flow.

Step 3 Adjust the torch acetylene flow.
- Open the acetylene cylinder valve slightly until the pressure registers and stabilizes on the acetylene high-pressure gauge; then open the valve about ½ turn.

> **CAUTION**
>
> Be sure to leave the valve wrench on the acetylene cylinder valve while the cylinder is in use so that the valve can be closed quickly in case of an emergency.

- Turn the acetylene regulator adjusting screw clockwise until the desired pressure shows on the acetylene working-pressure gauge.
- Open the acetylene valve on the torch handle.
- Check the regulator and ensure that the pressure remains at the desired setting. Adjust as needed.
- First close, and then open the torch acetylene valve about ½ turn to prepare for ignition.

> **WARNING!**
>
> When lighting the torch, be sure to wear flame-resistant gloves and goggles; use an appropriate device, such as a cup-type striker, to light the torch; and point the torch away from you. Always light the fuel gas first, and then open the oxygen valve on the torch handle.

Step 4 Light the torch.
- Hold the striker in one hand and the torch in the other hand. Strike a spark in front of the escaping acetylene gas.
- Open the acetylene valve on the torch until the flame jumps away from the tip about ¹⁄₁₆ inch (1.6 mm). Then slowly close the valve until the flame just returns to the tip (*Figure 26*). This sets the proper fuel gas flow rate for the size tip being used.

> **NOTE**
>
> It is common for visible soot to form in strings and float away from the acetylene-only flame. However, they can fall on an otherwise clean surface and be somewhat messy. Take the surroundings into account as you light the torch and be aware that soot is likely to form from the initial flame.

- Open the oxygen valve on the torch slowly to add oxygen to the burning acetylene. Observe the luminous cone at the tip of the nozzle and the long, greenish envelope around the flame, which is excess acetylene that represents a carburizing flame. As oxygen is added, the envelope of acetylene should disappear. The inner cone will appear soft and luminous, and the torch will make a soft, even, blowing sound. This indicates a neutral flame, which is the ideal flame for welding. If too much oxygen is added, the flame will become smaller, more pointed, and white in color, and the torch will make a sharp snapping or whistling sound. For brazing thin materials, a cooler, carburizing flame can be used to help prevent melting the base metal accidentally.

Step 5 Shut off the torch when finished brazing.
- Shut off the oxygen valve on the torch handle first.
- Shut off the acetylene valve on the torch handle. Close it quickly to avoid excess carbon buildup.
- Shut off both the oxygen and acetylene cylinder valves completely.
- Open both valves on the torch handle to release the pressure in the hoses. Watch the gauges on both regulators. With the cylinder valves closed, the pressure should fall on all regulator gauges until they register zero. Then close the valves on the torch handle.
- Turn the oxygen and acetylene regulator adjusting knobs counterclockwise (out) until loose. This will release the spring pressure on the diaphragms in the regulators.

Step 6 Properly stow the hoses while any final preparations to braze are made.

Note that many gas suppliers mount transducers on their tanks so that they can readily identify the tanks. The transducer is electronically scanned, and a coded number is matched against the supplier's records for identification. This aids in quickly determining the purchaser or user of the tank as well as the required retesting date. Acetylene cylinders are also equipped with safety plugs that will release if the temperature exceeds 220°F (104°C). The safety plugs release the gas in the event of a fire in order to prevent the cylinder

INITIAL FLAME

FINAL FLAME ADJUSTMENT

85204-13_F26.EPS

Figure 26 Reduce the acetylene flow until the flame returns to the tip.

from exploding. Note the transducer and safety plugs shown in *Figure 27*.

4.3.2 Lighting Air-Acetylene Equipment

Air-acetylene torches have limited applications, but may be used occasionally. Heat provided by an air-acetylene torch alone is suitable for brazing smaller lines. Air-acetylene torches are designed to draw in the correct amount of air for combustion from the atmosphere. Since air is roughly 80 percent nitrogen, the flame produced by an air-acetylene torch burns at a lower temperature than the flame produced by an oxy/acetylene torch. Some air-acetylene torches are self-igniting. On other torches the flame must be lit using a cup-type striker. Once the torch is lit it can be adjusted. Air-acetylene torches rely on flame action to pull in air, and are adjusted by adjusting the fuel control valve.

CYLINDER TOP SAFETY PLUGS (1 OF 2)

VALVE HANDWHEEL

IF PRESENT, GAS SUPPLIER TRANSDUCER FOR CYLINDER IDENTIFICATION

CYLINDER BOTTOM SAFETY PLUGS

85204-13_F27.EPS

Figure 27 Transducer, top and bottom safety plugs.

Air-acetylene torch tips are available in different sizes. However, it should be noted that a larger tip does not increase the temperature of the flame. It does, however, provide more heat through additional coverage of the material. Follow the manufacturer's guidelines for the selection of torch tips.

4.4.0 Brazing Joints

Once the flame is properly adjusted, the torch is ready to braze joints. When brazing rings are to be used, seat the ring at the bottom of the fitting using the tubing, or insert it into the groove provided. The split in the ring will allow it to slightly compress without deforming. Ensure that there is close contact between the end of the tube and the ring. Use the following procedure to braze a joint.

Step 1 Prepare the work area and protect any components in the piping system that can be damaged by the heat of the brazing process. Remove any loose, flammable materials in the area and ensure there is adequate room for physical movement around the joint. Ensure that an appropriate fire extinguisher is within easy reach.

Step 2 Put on tinted goggles with a minimum No. 3 tint (*Figure 28*).

Step 3 Set up and light the oxyacetylene brazing equipment as described previously.
  • Place the chosen filler metal within easy reach and be certain that the material is freely accessible while the process is in progress.
  • Light and adjust the torch to produce a neutral flame.

Step 4 Apply the heat to the tubing first, in the area adjacent to the fitting. Watch the flux (if used). It will first bubble and turn white and then melt into a clear liquid, becoming calm. At this time, shift the flame to the fitting and hold it there until any flux on the fitting turns clear.

Step 5 Continue to move the heat back and forth over the tubing and the fitting. Allow the fitting to receive more heat than the tubing by briefly pausing at the fitting while continuing to move the flame back and forth. Pause at the back of the fitting cup. For 1¼ inch (32 mm) and larger tubing, it may be difficult to bring the whole joint up to temperature at one time with a single-orifice tip. It will often be desirable to use a multi-flame (rosebud) heating tip such as the one shown in *Figure 29* to maintain a more uniform temperature

85204-13_F28.EPS

Figure 28 Typical welding goggles.

Figure 29 Multi-flame, or rosebud, heating tip.

over large areas. A mild preheating of the entire fitting and adjacent tubing is recommended for larger sizes, and the use of a second torch to retain a uniform preheating of the entire fitting assembly may be necessary with the largest diameters.

Step 6 As the joint is heated, the brazing ring will melt and begin to flow. No face-feeding of filler metal is allowed until a ring of filler metal from the brazing ring can be seen around at least two-thirds of the fitting perimeter. The joint will be considered defective if this stage is never

reached. If a ring is clearly visible around 100 percent of the fitting perimeter, face-feeding of filler metal is not required at all. However, it can be added if desired or specified.

Step 7 If needed, touch the filler metal rod to the joint. If the filler metal does not melt on contact, continue to heat and test the joint until the filler metal melts. Be careful to avoid melting the base metal.

Step 8 Hold the filler metal rod to the joint, and allow the filler metal to enter into the joint while holding the torch slightly ahead of the filler metal and directing most of the heat to the shoulder of the fitting. When the filler metal is applied to the top of the fitting, it should run down to the bottom and fill in the joint through capillary action. Applying heat at the bottom of the joint should help to make sure the filler metal penetrates. Make sure the filler metal is visible all around the shoulder of the joint. If it is not, apply additional filler metal. However, building an excessive **fillet** (extra shoulder of filler metal) does not improve joint quality and wastes material.

Step 9 For larger joints, small sections of the joint can be heated and brazed. Be sure to overlap the previously brazed section as brazing continues around the fitting (*Figure 30*).

Step 10 After the filler metal has hardened and cooled, wash the joint with a wet rag to clean any excess brazing flux from the joint. This also helps the copper oxides to crack and break away in flakes.

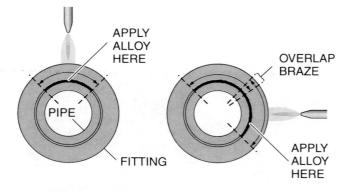
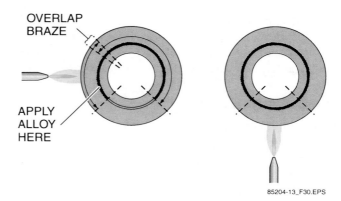

Figure 30 Working in overlapping sectors.

Allow the joint to cool below 250°F before applying a rag.

Step 11 Allow the joints to finish cooling naturally.

> **WARNING!**
>
> To avoid personal injury and joint damage, allow the joint to cool below 250°F (121°C) before quenching with water or performing any other form of rapid cooling.

The technique for brazing will vary somewhat, depending on the position in which the brazing is being done. The method just described is good for horizontal runs of pipe, but some alterations are needed for vertical-up and vertical-down brazing.

- *Vertical-up joints* – Heat the tube first, and then transfer the heat to the fitting. Move the heat back and forth between the fitting and the tube, but be sure not to overheat the tube below the fitting. Doing so could cause the filler metal to flow out of the joint. When the brazing temperature is reached, apply the filler metal to the joint while applying heat to the wall of the fitting. This should cause the filler metal to run up into the fitting. Heat and capillary action must be used to defeat gravity.
- *Vertical-down joints* – Heat the tube before heating the fitting. When the brazing temperature is reached, apply more heat to the fitting while applying the filler metal to the joint. The filler metal should run down into the fitting.

Remember that, when a joint with a brazing ring has been sufficiently heated, there should be a line of filler metal around the perimeter of the finished joint. The presence of this line confirms that the brazing ring filler metal has completely penetrated the joint. Some specs do still require the addition of a bead of filler metal around the fitting to cap the joint, even when a full ring is visible.

4.4.1 Brazing Metals Other Than Copper

Sometimes the maritime pipefitter will be asked to braze fittings made of something other than copper. Typical metals that may have to be brazed include stainless steel, brass, nickel-copper alloys (sometimes called cupronickel), valve bronze, and Inconel (an alloy of nickel and chromium). Brazing temperatures are similar to those for copper

brazing, with some exceptions. Some special filler considerations for these metals are as follows:

- *Stainless steel* – Filler metals containing either cadmium or zinc can cause excessive corrosion. To avoid damage to the stainless steel, low temperature silver brazes are usually used. If the fitting will be installed in a highly corrosive area, it may be necessary to use special filler metals containing chemically inert metals, sometimes called noble metals. Noble metals are the least prone to corrosion, and will not promote corrosion on stainless steel at the weld area.
- *Brass* – When brazing in an enclosed area, use a non-matching filler metal, such as aluminum bronze or silicon bronze, to reduce toxic fume formation. In open areas, silver-based filler brazing with matching flux creates a satisfactory joint.
- *Valve bronze* – Bronze melting points vary depending on the tin content, and some bronze has a melting point as low as 950°F (510°C). Be careful to pick a type of filler metal that has a lower melting point than the fitting.
- *Inconel* – Use special nickel alloy filler, in the *AWS A5.14* series, to braze Inconel fittings.

In most cases of brazing metals other than copper, standard fluxes suitable for copper can be used. AWS-type flux is recommended for some stainless steel brazing. Flux-coated rods are often used. When brazing rings are specified, they must be of the proper material for the metal being brazed.

When brazing copper pipes, a neutral flame usually produces the best bond. This is not the case with other metals and the flame should be adjusted as explained in the following list:

- Stainless steel brazing requires a slightly reducing flame.
- For brass, use a slightly oxidizing flame.
- To braze nickel-copper/cupronickel use a reducing flame.
- For bronzes, use a slightly oxidizing flame.
- When brazing Inconel use a slightly reducing flame.

4.4.2 Brazing Dissimilar Metals

Sometimes the maritime pipefitter will be called on to braze two dissimilar metals. Pay special attention to these situations. Flame temperature, filler metal, and flux type should be carefully selected.

For example, when brazing dissimilar metals such as copper to brass or copper to steel, flux and a silver-bearing brazing alloy such as a BCuP-4 should be used. Remember that, for copper-to-copper joints, a phosphorous-bearing filler metal acts as a flux, but additional flux is almost always required or specified. When a non-phosphorous filler metal is used, a flux must be used, even on copper. As always, the pieces being joined must be very clean and free of any oil or grease. Aside from choosing the correct filler metal and flux, the other important aspect of brazing dissimilar metal is the application of heat. The fittings on accessories that use steel or copper clad steel fittings reach their melting point faster than copper, so there is a potential to overheat and distort the fittings during brazing. In order to avoid overheating, it is necessary to move the heat rapidly and uniformly around the joint, and not let it linger too long in one place. Thin steel can turn cherry-red very quickly, since it transfers heat away much more slowly than copper, allowing its own temperature to rise quickly. When joining copper to steel, more heat needs to be applied to the copper, since copper is the better conductor and thus will dissipate heat more quickly than the steel.

Also keep in mind that metals expand at different rates. Copper and brass will expand more than steel when heated. This may cause the joint to tighten up so much that the filler metal will not flow into the joint. For that reason, a joint clearance of 0.010 inch (0.26 mm) may be more suitable than the usual 0.001 to 0.003 inch (0.026 to 0.076 mm) clearance when brazing copper or brass to steel. However, if the heat is likely to loosen the fit between the two metals, start with a tighter fit.

SUMMARY

Brazing uses filler metals that melt at temperatures above 842°F. Acetylene, or acetylene combined with oxygen, is generally used for brazing. The air-acetylene method is commonly used for smaller-diameter tubing. The oxyacetylene method generates a much greater heat than the air-acetylene method. Joining is faster using the oxyacetylene method and thus results in less oxidation of the joint.

Brazing procedures require some special tools. These include setups for air-acetylene or oxyacetylene torches, including tanks, pressure regulators, hoses, torch handles, and specific brazing tips. Cleaning equipment and reamers are also needed. Common brazing materials are filler metals and fluxes. Many types of filler metals and fluxes are available to match specific brazing needs.

Brazing procedures require a step-by-step approach that includes measuring and cutting new pipe, then cleaning and applying flux to the joints. After this is done, the joints should be aligned. Then the joints can be heated to the correct temperature. The next step is bringing the nonferrous filler metal into contact with the tubing where it melts and flows into the joint by capillary action. Once a solid, leak-proof joint is obtained, it can be cleaned and allowed to cool.

Special temperature considerations, filler materials, and fluxes are sometimes needed when brazing pipes made of other metals than copper, and when joining two dissimilar metals.

The use of high heat, flame, and flammable gases creates unique hazards for workers performing brazing work. Acetylene is volatile and must be carefully handled. Oxygen is flammable and can explode if it comes into contact with oil or grease with a source of ignition. There are special safety procedures for handling, storing, and transporting oxygen and acetylene cylinders that must be followed.

1. Capillary action is the tendency of a liquid to flow between _____.

   a. corroded spaces
   b. wide spaces
   c. narrow spaces
   d. clean spaces

2. An oxyacetylene torch produces a hotter flame than an air-acetylene torch.

   a. True
   b. False

3. The acetylene cylinder contains acetylene in solution with _____.

   a. acetone
   b. helium
   c. oxygen
   d. water

4. An acetylene regulator usually has _____.

   a. one pressure gauge
   b. two pressure gauges
   c. three pressure gauges
   d. four pressure gauges

5. The size of a brazing nozzle tip opening can be determined by _____.

   a. measuring the opening with a wire gauge
   b. counting the number of holes in the tip
   c. measuring the length of the tip from the handle pivot
   d. observing the tip number

6. A nonferrous filler metal contains no _____.

   a. iron
   b. copper
   c. silver
   d. brass

7. Which of the following pipe preparation steps should be done *first*?

   a. Clean the fitting ends.
   b. Measure the difference between the faces of the two fittings.
   c. Ream the tubing ends.
   d. Cut the tubing to length

8. To clean the interiors of fittings before brazing, use _____.

   a. sandpaper
   b. emery cloth
   c. reamer
   d. a brush

9. A hacksaw should be used to cut tubing in most maritime applications.

   a. True
   b. False

10. Which of the following could contaminate freshly cleaned fittings?

    a. The flux paste brush
    b. The tubing cutter
    c. Your fingers
    d. Your breath

11. Acetylene equipment is usually used for brazing because acetylene _____.

    a. is cheaper than other gases
    b. is less toxic than other gases
    c. is the only gas that will mix with oxygen
    d. provides enough heat to give good control of the brazing process

12. Acetylene and oxygen cylinders should be stored without a regulator attached.

    a. True
    b. False

13. Using pressure to remove debris from the acetylene and oxygen hoses is called _____.

    a. pressurizing
    b. blowing down
    c. clearing
    d. purging

14. When first lighting the oxyacetylene torch, the oxygen valve on the torch handle should be _____.

    a. fully open
    b. completely closed
    c. ¼ turn open
    d. ½ turn open

15. When brazing large fittings, it is permissible to braze small sections of the fitting, as long as the brazed sections are _____.

    a. cooled quickly
    b. overlapped
    c. re-melted at least three times
    d. not copper

Trade Terms Introduced in This Module

Acetone: A colorless organic solvent that is volatile and extremely flammable.

Alloy: Any substance made up of two or more metals.

Brazing ring: Filler alloy shaped in a circle for insertion in a fitting. Brazing rings are used instead of conventional filler metals in certain situations.

Brazing: A method of joining metals with a nonferrous filler metal using heat above 842°F but below the melting point of the base metals being joined; also incorrectly known as hard soldering.

Capillary action: The tendency of a liquid to flow between narrow spaces. Capillary action is the gripping of a liquid to the walls of its container, combined with surface tension holding the surface of the liquid intact.

Cup depth: The distance that a tube inserts into a fitting, usually determined by a stop inside the fitting.

Fillet: A rounded internal corner or shoulder of filler metal. Often appears at the meeting point of a piece of tubing and a fitting when the joint is soldered or brazed.

Flashback arrestor: A valve that prevents a flame from traveling back from the tip and into the hoses.

Flux: A chemical substance that prevents oxides from forming on the surface of metals as they are heated for soldering, brazing, or welding.

Nonferrous: A group of metals and metal alloys that contain no iron.

Oxidation: The process by which the oxygen in the air combines with metal to produce tarnish and rust.

Wetting: A process that reduces the surface tension so that molten (liquid) filler flows evenly throughout the joint.

Additional Resources

This module presents thorough resources for task training. The following resource material is suggested further study.

A Guide to Brazing and Soldering. The Harris Products Group. Available at **www.harrisproductsgroup.com**.
Gas Welding, Cutting, Brazing, & Heating Torch Instruction Manual. The Harris Products Group. Available at **www.harrisproductsgroup.com**.
www. brazingbook.com. Lucas-Milhaupt, Inc. An interactive resource on the subject of brazing.

Figure Credits

NCCER CURRICULA — USER UPDATE

NCCER makes every effort to keep its textbooks up-to-date and free of technical errors. We appreciate your help in this process. If you find an error, a typographical mistake, or an inaccuracy in NCCER's curricula, please fill out this form (or a photocopy), or complete the online form at **www.nccer.org/olf**. Be sure to include the exact module ID number, page number, a detailed description, and your recommended correction. Your input will be brought to the attention of the Authoring Team. Thank you for your assistance.

Instructors – If you have an idea for improving this textbook, or have found that additional materials were necessary to teach this module effectively, please let us know so that we may present your suggestions to the Authoring Team.

NCCER Product Development and Revision

13614 Progress Blvd., Alachua, FL 32615

Email: curriculum@nccer.org
Online: www.nccer.org/olf

❏ Trainee Guide   ❏ Lesson Plans   ❏ Exam   ❏ PowerPoints   Other _____

Craft / Level: _____   Copyright Date: _____

Module ID Number / Title: _____

Section Number(s): _____

Description: _____

_____

_____

_____

_____

Recommended Correction: _____

_____

_____

_____

_____

Your Name: _____

Address: _____

_____

Email: _____   Phone: _____

85205-13

Threaded Pipe Fabrication

Module Five

Trainees with successful module completions may be eligible for credentialing through NCCER's National Registry. To learn more, go to **www.nccer.org** or contact us at **1.888.622.3720**. Our website has information on the latest product releases and training, as well as online versions of our *Cornerstone* magazine and Pearson's product catalog.

Your feedback is welcome. You may email your comments to **curriculum@nccer.org**, send general comments and inquiries to **info@nccer.org**, or fill in the User Update form at the back of this module.

This information is general in nature and intended for training purposes only. Actual performance of activities described in this manual requires compliance with all applicable operating, service, maintenance, and safety procedures under the direction of qualified personnel. References in this manual to patented or proprietary devices do not constitute a recommendation of their use.

Objectives

When you have completed this module, you will be able to do the following:

1. Identify and describe characteristics of pipe threads and fittings.
2. Calculate pipe lengths between threaded joints.
3. Describe how to thread and assemble threaded pipe and fittings.

Performance Tasks

Under the supervision of the instructor, you should be able to do the following:

1. Determine pipe lengths between fittings using the center-to-center method.
2. Determine pipe lengths between fittings using the center-to-face method.
3. Determine pipe lengths between fittings using the face-to-face method.
4. Assemble threaded pipe using various fittings.

Trade Terms

Banded fitting
Cast iron
Ductile
Forged steel
Galling
Length of effective thread
Makeup

Malleable iron
National Pipe Thread (NPT)
Takeout
Thread angle
Thread crest
Thread engagement

Industry-Recognized Credentials

If you are training through an NCCER-accredited sponsor, you may be eligible for credentials from NCCER's Registry. The ID number for this module is 85205-13. Note that this module may have been used in other NCCER curricula and may apply to other level completions. Contact NCCER's Registry at 888.622.3720 or go to **www.nccer.org** for more information.

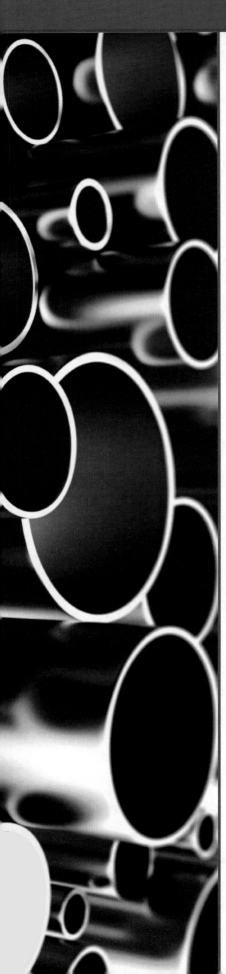

Contents

Topics to be presented in this module include:

1.0.0 Introduction...1

2.0.0 Pipe Threads 1

    2.1.0 Types of Threads..1

        2.1.1 NPT Threads 1

    2.2.0 The Tapered Thread Connection 2

3.0.0 Pipe Fittings 3

    3.1.0 Forged Steel Fittings...3

    3.2.0 Elbows, Offsets, and Return Bends...4

    3.3.0 Branch Connections...4

    3.4.0 Caps and Plugs 6

    3.5.0 Line Connections...6

        3.5.1 Couplings and Reducing Couplings..6

        3.5.2 Unions 7

        3.5.3 Reducing Bushings..7

    3.6.0 Nipples 7

    3.7.0 Flanges 8

4.0.0 Determining Pipe Lengths Between Fittings 8

    4.1.0 Center-to-Center Method 9

    4.2.0 Center-to-Face Method 12

    4.3.0 Face-to-Face Method 13

5.0.0 Threaded Pipe Assembly..13

    5.1.0 Pipe Joint Tape and Compounds 13

        5.1.1 PTFE Tape 13

        5.1.2 Pipe Joint Compounds..14

    5.2.0 Selecting Wrenches 15

    5.3.0 Fitting Screwed Pipe and Fittings...16

    5.4.0 Installing Threaded Valves 16

Figures and Tables

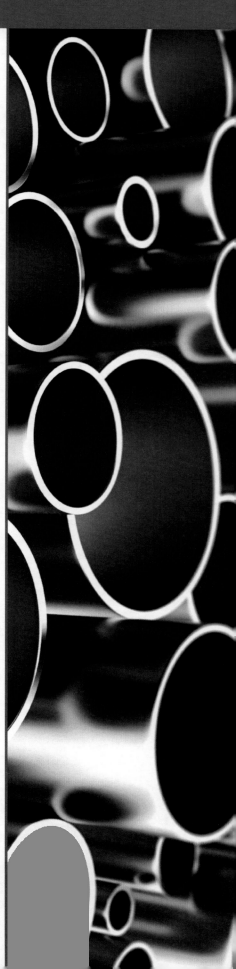

Figure 1   NPT pipe thread details 2
Figure 2   Perfect and imperfect threads 2
Figure 3   Threaded pipe fittings 4
Figure 4   Elbows 5
Figure 5   Street elbow 5
Figure 6   Offset 5
Figure 7   Return bends 5
Figure 8   Branch connections 6
Figure 9   Caps and plugs 7
Figure 10 Standard coupling 7
Figure 11 Concentric and eccentric reducing couplings 7
Figure 12 Union 8
Figure 13 Dielectric union 8
Figure 14 Reducing bushing, or fitting reducer 8
Figure 15 Types of nipples 9
Figure 16 Threaded flange 9
Figure 17 Raised- and flat-faced flanges 9
Figure 18 Takeout and makeup dimensions 10
Figure 19 Center-to-center pipe measurement 10
Figure 20 Center-to-face dimension of a 2,000-pound 90-degree ell 10
Figure 21 Center-to-center method of determining pipe length 11
Figure 22 Center-to-face method of determining pipe length 12
Figure 23 Face-to-face measurement between two fittings 13
Figure 24 White, yellow, and pink PTFE tape 14
Figure 25 PTFE tape and pipe joint compound applied 14
Figure 26 Pipe joint compounds 15
Figure 27 Proper thread makeup for a valve 17

Table 1   American National Standard Taper Pipe Thread
           (NPT) Dimensions 3
Table 2   Example of Makeup Chart 11
Table 3   Suggested Wrench-to-Pipe Sizes 15

1.0.0 INTRODUCTION

The ability to install threaded pipe in accordance with the job requirements and specifications is a skill every pipefitter must develop. Threaded connections are relatively inexpensive to fabricate and a common way to join pipe. Throughout a maritime pipefitting career, various threaded piping systems will be encountered. The purpose of this module is to introduce the maritime pipefitter trainee to the processes and procedures used to thread and assemble pipe.

There are several advantages to a threaded piping system. These systems require no specialized in-shop fabrication and can be easily fabricated on the job site using pipe and various fittings. Portable power and manual threading equipment can be used almost anywhere on the job site. If a piping system must be installed near flammable liquids or gases, threaded pipe is much safer and faster than welded systems because fire hazards can be avoided. Threaded pipe can be cleaned on the inside before it is installed, reducing the possibility of metal particles becoming entrapped and damaging valves or strainers. Debris is more likely to enter welded piping systems during the welding process. Finally, threaded piping systems are easier to disassemble and access for component and piping repairs.

Threaded pipe also has its disadvantages. Threaded pipe joints are more prone to leak than welded joints. In addition, the strength of the pipe is reduced by the threading process because the wall thickness of the pipe is reduced at the threads. Threaded pipe cannot be used if severe erosion, crevice corrosion, extreme shock, or consistent vibration is anticipated. However, the weaknesses caused by the threading process are minimized when pipe weights higher then Schedule 40 are used.

In this module, you will be exposed to the characteristics of pipe threads and the related fittings; methods to accurately calculate threaded pipe lengths; and how to properly thread and assemble piping system components.

2.0.0 PIPE THREADS

In technical terms, a pipe thread is an inclined plane wrapped around a cylinder. A pipe thread can also be described as a continuous groove that spirals around a pipe or other material. The threads provide a means of firmly attaching components together. For the attachment to be made, one part must have a thread on the exterior surface while the other has a thread on an interior surface. These are referred to as male and female threads, respectively.

The **thread angle** is the angle formed by the two inclined faces of the thread. The thread angle can vary depending on the specific thread standard used. The thread pitch refers to the distance between adjacent **thread crests**. This distance determines how many threads there are in one inch of thread length, spoken as threads per inch, For example, eight threads per inch is a typical value for larger pipe sizes. Many pipefitters will refer to the value of eight threads per inch as the thread pitch. The root of the thread is the innermost part of the thread when viewed from the side.

2.1.0 Types of Threads

The **National Pipe Thread (NPT)** is by far the most common thread standard for pipe connections. However, there are also British Standard Pipe Parallel (BSPP) and British Standard Pipe Tapered (BSPT) threads. BSPP threads are also known as Whitworth threads. In the US, there is another pipe thread standard known as National Pipe Straight (NPS). Under this standard, the threaded portion of the pipe has no taper at all, making it appear more like the straight threads on a common bolt. However, the tapered characteristic of the NPT is very significant in its ability to positively seal against leakage, as well as its ease of assembly. Once tightened, the tapered threads also help to form a more rigid structure. NPS threads are not used in piping systems, since they are unable to form a positive seal. They are most often used for simple assembly applications. Therefore, the discussion of threading in this module will focus solely on the NPT type.

2.1.1 NPT Threads

The details of the NPT standard are shown in *Figure 1*. The NPT has a thread angle of 60 degrees. Both the crest and the root of the thread are very slightly flattened. The taper angle used results in a taper of $\frac{1}{16}$-inch per inch of threads ($\frac{1}{32}$-inch on each side of the pipe).

When a thread is created on a pipe, there are about seven to twelve perfect threads and several imperfect threads (*Figure 2*). Imperfect threads are those that are not cut to their full depth as a result of the taper, as opposed to perfect threads. The actual number of perfect threads created depends on the size of the pipe being threaded. It is important to note that the term *perfect threads* relates only to the fact that they are cut to the full depth. As shown in *Figure 2*, the threads closest to the end of the pipe are perfect threads; they are cut to the full depth and the crests are properly formed. The remaining threads are imperfect because they are not completely cut, resulting in rounded, or imperfect crests. The primary sealing power of the threads occurs as the female threads begin to meet the tapered imperfect threads. If the perfect threads are marred or damaged, they will lose their sealing power.

Threads are designated by specifying in sequence the nominal pipe size, the thread pitch, and the thread standard symbols.

For example, the thread specification ¾ – 14 NPT means:

¾ = ¾" nominal pipe size, ID
14 = 14 threads per inch (pitch)
NPT = National Pipe Thread

It should be noted that pipe threads are also referred to with a number of other acronyms, both in print and in conversation. The acronyms include the following:

- MIP (Male Iron Pipe)
- FIP (Female Iron Pipe)
- IPT (Iron Pipe Thread)
- FPT (Female Pipe Thread)
- MPT (Male Pipe Thread)

The MIP, MPT, FIP, and FPT acronyms are particularly useful when selecting pipe fittings and piping system components. They designate whether a component is male or female so that installers know what gender the mating portion must be.

2.2.0 The Tapered Thread Connection

In order to have a strong, leak-proof threaded joint, the threads must be clean and smoothly cut.

They must have the correct pitch, taper, and form. If these conditions are not met, the sealing surfaces will leak or the joint cannot be assembled.

Tapered pipe threads are engaged or made up in two phases: hand-tight engagement and wrench makeup. Hand engagement refers to how far the fitting can be tightened by hand. Wrench makeup is the additional turning of the fitting with a wrench to completely tighten the joint. *Table 1* shows the dimensions for hand-tight engagement as well as other NPT specifications for commonly used pipe sizes. In practice, about four to five turns are done by hand, followed by several more turns with a wrench. A greater number of turns are required for larger pipe. When a pipe is threaded and properly tightened, about three threads (most if not all imperfect) should remain showing.

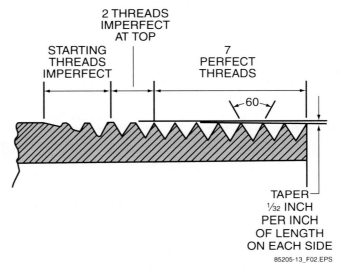

Figure 2 Perfect and imperfect threads.

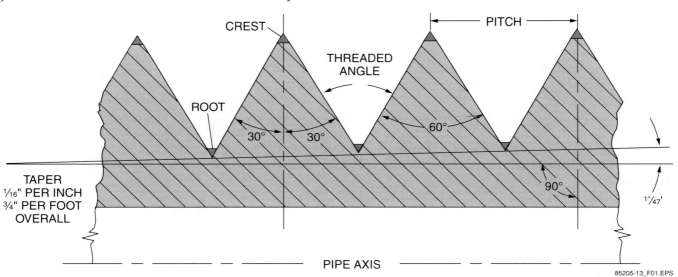

Figure 1 NPT pipe details.

Table 1 American National Standard Taper Pipe Thread (NPT) Dimensions.

Nominal Pipe Size (inches)	Threads per Inch	No. of Usable Threads	Hand-Tight Engagement (inches)	Total Thread Makeup (inches)	Total Thread Length (inches)
1/8	27	7	3/16	1/4	3/8
1/4	18	7	1/4	3/8	5/8
3/8	18	7	1/4	3/8	5/8
1/2	14	7	5/16	7/16	3/4
3/4	14	8	5/16	1/2	3/4
1	11 1/2	8	3/8	9/16	7/8
1 1/4	11 1/2	8	7/16	9/16	1
1 1/2	11 1/2	8	7/16	9/16	1
2	11 1/2	9	7/16	5/8	1
2 1/2	8	9	11/16	7/8	1 1/2
3	8	10	3/4	1	1 1/2
3 1/2	8	10	13/16	11/16	1 5/8
4	8	10	13/16	11/16	1 5/8
5	8	11	15/16	13/16	1 3/4
6	8	12	15/16	13/16	1 3/4

Hand engagement provides an opportunity to check the quality of threads after cutting. Poorly formed pipe threads will be difficult to engage properly by hand. Snags or rough spots can be felt through the hand that would not likely be detected when a wrench is applied.

It is important to note that the overall quality of a thread may be dictated by the job specifications. Specific testing procedures may be required to ensure that threads meet a minimum standard of dimensions and tolerances.

3.0.0 PIPE FITTINGS

Threaded piping systems can consist of any number of different components connected by pipe. The individual sections of pipe are connected by various fittings. Fittings may also be needed to make the final connection to a component. Pipe fittings come in various sizes, materials, strengths, and designs to match the various piping systems and can be either plain fittings or **banded fittings**. The banded fittings are malleable and **cast iron** fittings, while other fittings are made from **forged steel**. The **malleable iron** fittings are banded to provide extra strength. Pipe fittings provide the means of changing the direction of flow, connecting a branch line to a main line, closing off the end of a line, or joining two pipes together of the same or different sizes. Pipe fittings can be grouped according to their use:

- Elbows, offsets, and return bends

- Branch connections
- Caps and plugs
- Line connections

Pipe fittings are primarily classified by their materials of construction and a pressure class, which allows the user to determine a fitting's ability to withstand pressure.

3.1.0 Forged Steel Fittings

It is important to note that malleable and cast iron threaded fittings are rarely used in the maritime environment. The specifications for a number of clients and projects prohibit their use, since they are not as strong and durable as their forged steel counterparts are. Although maritime pipefitters should be somewhat familiar with them and some will be pictured in this module, forged steel fittings will be used for most maritime applications. The process of constructing a threaded piping run or system is the same, regardless of the fitting material.

Cast iron fittings are made by heating the iron and carbon alloy to its liquid state, and then pouring the material into a cast where it cools and hardens. Threaded fittings are typically available only up to a pressure class of 250 psi. They are somewhat brittle, and tend to crack under stress. Cast fittings typically have a visible seam on the exterior, where the two halves of the casting mold meet.

Malleable iron is derived from cast iron and has less carbon content. Malleable iron is more **ductile** and less likely to crack under stress. Malleable iron fittings are available up to a pressure class of 300 psi.

Forged steel fittings are far more durable than either cast or malleable fittings. Forged steel fittings are made by heating the steel to a high temperature, but without reaching the liquid state. This state may be referred to as its plastic state. The hot steel is then pressed, hammered, or rolled into the desired shape. Forged steel fittings are available in pressure rating classes of 2,000 psi, 3,000 psi, and 6,000 psi. *Figure 3* shows common fittings used on threaded pipe.

Regardless of the material a threaded fitting is made from, the threads are machined into the fitting after it is formed. Threads are not created on the fitting during the casting or forging stages.

3.2.0 Elbows, Offsets, and Return Bends

Elbows and return bends are fittings used for changing the direction of fluid flow in a piping system. An elbow, commonly referred to as an ell, an L, a 90, or a 45, is a fitting that forms an angle between different connecting pipes. This angle is usually 90 degrees, although 45-degree ells are also commonly used for offsets. Elbows come in several different angles. The 11¼-, 5⅝-, and 30-degree elbows are rarely used and very hard to obtain. The 90- and 45-degree elbows are by far the most common. *Figure 4* shows 5⅝-, 11¼-, 22½, 30-, 45-, and 90-degree elbows. Reducing elbows can also be found, allowing two different sizes of pipe to be connected. Elbows can have male threads on one end and female threads on the other. This style, called a street ell, is useful to connect a pipe to a component at a 90-degree angle. *Figure 5* shows an example of a street elbow.

An offset changes the direction of flow for a short distance. The offset moves the pipe run to one side and then returns it to the same direction. An offset can be made by using two fittings, one at each end of the offset. *Figure 6* shows an example of an offset.

A return bend is a U-shaped fitting that sends the fluid back in the same direction from which it came. Return bends are often used in boilers, radiators, and systems in which the pipe must pass through the same area several times. Return bends are manufactured in various sizes and come in close, open, and wide patterns. *Figure 7* shows examples of return bends.

3.3.0 Branch Connections

Branch connections often divide the flow of a fluid and send it in two different directions.

Figure 3 Threaded pipe fittings.

85205-13_F03.EPS

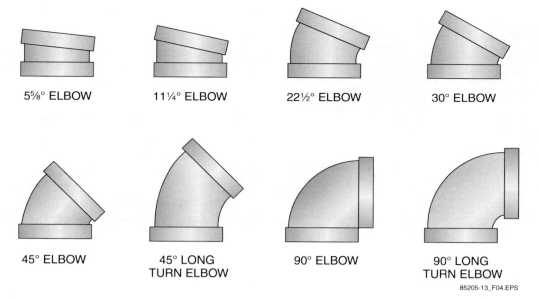

5⅝° ELBOW   11¼° ELBOW   22½° ELBOW   30° ELBOW

45° ELBOW   45° LONG TURN ELBOW   90° ELBOW   90° LONG TURN ELBOW

85205-13_F04.EPS

Figure 4 Elbows.

85205-13_F05.EPS

Figure 5 Street elbow.

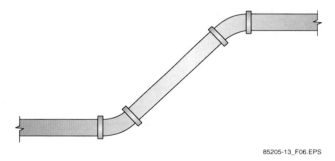

85205-13_F06.EPS

Figure 6 Offset.

Branch connections are also used to merge two separate streams of flow into a single stream. Tees are the most widely used of the branch connections. They are used to provide a 90-degree branch connection to another pipe. Reducing tees, used to connect lines of different sizes as well as providing a branch connection, are also used. Tees that accommodate a different pipe size at all three connections are available, but are harder to acquire. A cross provides a point of connection for four separate pipes. Both tees and crosses can be reducing types, where the pipe size of one or

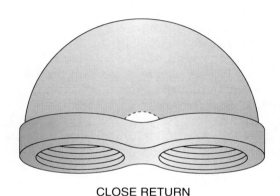

CLOSE RETURN

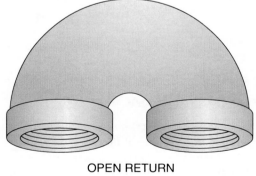

OPEN RETURN

85205-13_F07.EPS

Figure 7 Return bends.

more connections is smaller than the other(s). *Figure 8* shows common tee and cross fittings.

To help distinguish one connection from another on a tee fitting, they are given different names. The two connections that are directly in line with each other are referred to as the main or run connections. The third connection, placed

CROSS   TEE

85205-13_F08.EPS

Figure 8 Branch connections.

at a 90-degree angle from the other two, is called the branch.

3.4.0 Caps and Plugs

Caps and plugs (*Figure 9*) are used to close off the ends of pipes or other fittings. They have the same pressure ratings as other fittings. A cap screws onto a pipe end or the male end of a fitting, while a plug screws into a female fitting. Caps can be either plain with round heads, installed with a pipe wrench, or have a square projection on the body to allow for tightening with a box or adjustable wrench. Plugs can have solid square heads or hex-shaped heads for use with the proper wrench. They may also be countersunk, having a square or hex depression into which a square or hex key can be inserted for tightening. A bull plug is a long shaft that is threaded on one end that may or may not have a hex head. Bull plugs are used to plug an insulated pipe line so that the head will extend out of the insulation for easy removal. A nipple with a cap can also serve this same purpose.

3.5.0 Line Connections

Threaded couplings and unions typically join pipes of the same size. Reducing bushings (often called fitting reducers) and reducing couplings join pipes of different sizes.

3.5.1 Couplings and Reducing Couplings

A coupling is a sleeve that is used to connect two straight pieces of pipe. Both ends of a straight coupling are tapped with right-hand threads. Usually, all lengths of threaded pipe less than 8 inches in diameter are provided by the manufacturer with a straight coupling attached on one end. Of course, the coupling provided is not necessarily the type specified for the job at hand.

Figure 10 shows a standard coupling. A reducing coupling, also referred to as a bell reducer, is used to join the male ends of two different sizes of pipe.

Reducing couplings can be made in concentric and eccentric styles (*Figure 11*). An eccentric fitting offsets the center lines of the two connections. This arrangement allows the outside wall of each pipe to remain aligned on one side. Eccentric styles are typically malleable rather than forged however.

The problem with couplings is that they do not allow for easy disassembly of the line when repairs or changes are required. To provide an easy access point, a union can be used at convenient locations. For example, unions are often found near the final threaded piping connection to a component.

Smooth Flow

All piping systems resist the flow of fluids to some extent. This resistance originates from the friction of liquid or gaseous fluids moving along the pipe walls, rubbing against them. Any time the direction of flow is changed, additional friction is generated as the fluid presses harder against the pipe wall, physically pushed in a new direction. Imagine holding a garden hose with a jet nozzle on the end, with the water stream directed at a nearby wall. When the stream is directed at some angle to the wall, it is forced to change its direction, and friction between the water and the wall is a consequence. Left in place long enough (months or years perhaps), the friction of the stream will cause the wall to erode. The amount of friction increases as the angle of the water stream becomes closer to perpendicular. Conversely, the friction decreases as the angle of the water stream comes closer to being parallel with the wall.

When fittings are used to change the direction of flow in a piping system, the 90-degree elbow typically creates the greatest amount of friction. For this reason, 45-degree ells are used when possible or practical to create offsets and directional changes. The friction, and thus the resistance to fluid flow, is much lower when 45-degree ells are used. While using elbows of a lesser angle results in even less friction, ells at angles less than 45 degrees can be very hard to find and quite expensive.

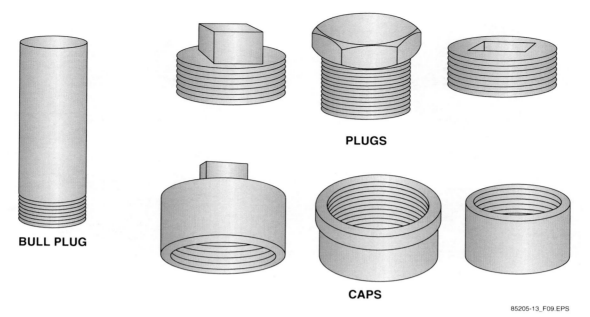

PLUGS

BULL PLUG

CAPS

85205-13_F09.EPS

Figure 9 Caps and plugs.

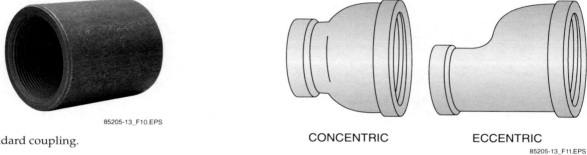

85205-13_F10.EPS

Figure 10 Standard coupling.

CONCENTRIC   ECCENTRIC

85205-13_F11.EPS

Figure 11 Concentric and eccentric reducing couplings.

3.5.2 Unions

Unions are used either to join two ends of pipe that cannot be turned or to permit the disconnecting of pipes at some future time without cutting them. A union usually consists of three parts: two sleeves that are threaded on the ends of the two pipes, and a threaded coupling ring to draw the two sleeves together. The male side of the union goes in the direction of the flow. *Figure 12* shows a union, along with a drawing to show the sealing surfaces inside. Unions often have different metals inserted into the assembly, such as brass, to create a better sealing surface. Others may also have a soft gasket or O-ring.

One special type of union is the dielectric union (*Figure 13*). It is used to prevent the breakdown of metals caused by directly connecting two different metals, such as a bronze valve and steel pipe. The dielectric union has an insulating gasket and ring between the two sleeves, preventing the surfaces from making physical contact. Connecting dissimilar materials together, combined with the flow of fluids through the connection, can often cause serious and rapid corrosion to occur.

3.5.3 Reducing Bushings

The function of a reducing bushing (*Figure 14*) is to connect the male end of a pipe to the female end of a fitting that is a larger size. It is also referred to as a fitting reducer. It consists of a hollow plug with male and female threads to suit the different diameters. Reducing bushings are usually made with a hexagon top but can also be made without the head. Those without a head are known as face bushings or flush bushings and can be screwed into the fitting to form a neat, flush finish.

3.6.0 Nipples

Nipples are merely short pieces of pipe threaded on both ends and used to make close connections between fittings. They are made in all sizes and types of pipe and are stocked in various lengths classified as close, shoulder, short, and long. *Figure 15* shows various types of nipples. The longest nipples are 12 inches. Nipple lengths up to 6 inches are considered short nipples; nipples

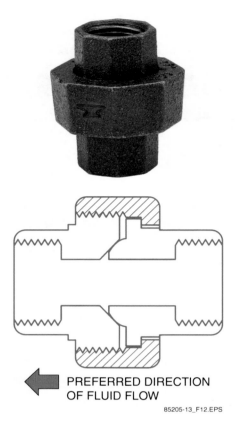

PREFERRED DIRECTION
OF FLUID FLOW

85205-13_F12.EPS

Figure 12 Union.

85205-13_F13.EPS

Figure 13 Dielectric union.

85205-13_F14.EPS

Figure 14 Reducing bushing, or fitting reducer.

between 6 and 12 inches are long nipples. Pipe lengths longer than 12 inches are known as cut pipe. A close nipple is threaded end to end, with a very small separation between the two threaded portions. A shoulder nipple has a slightly longer section of unthreaded pipe in the center.

> **NOTE**
>
> The abbreviations TBE (thread both ends) and TOE (thread one end) are used in instructions for fabrication of nipples.

The stated length of a nipple is its total length from end-to-end, including the threaded portion.

3.7.0 Flanges

The threaded flange is considered the weakest type of flange because the threads cut into the pipe. It is normally used only for low-pressure lines and pipes. As in all flanges, the bolt holes must be carefully aligned when making up the fitting. *Figure 16* shows a threaded flange.

Each type of flange is available in different face styles, and each flange face style uses a certain gasket. A set of two flanges of the same face style, called companion flanges, is used with the proper gasket installed between the two faces. Two com-

panion flanges can also be used on each side of a component such as a valve.

The two face styles of threaded flanges are the flat-faced flange and the raised-face flange. The flat-faced flange is flat and smooth across the entire face. This flange uses a gasket that covers the entire flange face and has bolt holes cut into it. The raised-face flange has a wide, raised rim around the center portion of the flange. Both raised- and flat-faced flanges typically have grooves or ridges on the raised portion. Since there is less surface area available as a sealing surface, the ridges are added to improve the seal where the flange meets the gasket. The portion of the two flanges with the bolt holes does not make contact when they are assembled. This flange uses a flat ring gasket that is squeezed tightly between the two raised faces. Bolt holes are not required in the gasket as a result. *Figure 17* shows the two flanges.

4.0.0 DETERMINING PIPE LENGTHS BETWEEN FITTINGS

Piping drawings usually show the rough dimensions of an overall run of pipe, but it is up to the pipefitter to field-fit the individual straight lengths. In order to install a piece of pipe

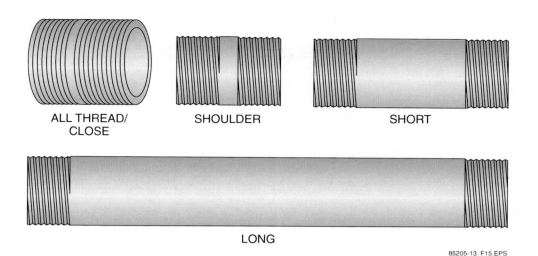

ALL THREAD/
CLOSE

SHOULDER

SHORT

LONG

85205-13_F15.EPS

Figure 15 Types of nipples.

85205-13_F16.EPS

Figure 16 Threaded flange.

between two fittings or connections, the pipefitter must be able to determine the **takeout** of the fittings to find the length of pipe to cut. Takeout refers to the distance from the face or shoulder of a fitting to the center of the fitting. Another important term used in determining pipe lengths is **makeup**, also known as the **thread engagement** or the **length of effective thread**. The makeup is basically the distance that the pipe intrudes into the fitting once it is tightened.

Figure 18 shows a tee fitting with connected pipe to demonstrate these terms. Knowing the proper takeout for the fittings is essential when determining the lengths of straight pipe. Three of the most common methods for determining pipe lengths are the center-to-center method, the center-to-face method, and the face-to-face method.

4.1.0 Center-to-Center Method

Pipe drawings may give the center-to-center dimensions for a section of pipe connected to fittings or components on each end. *Figure 19* is a simple pipe drawing showing a center-to-center measurement between two fittings. The challenge is to determine the exact length to which the pipe must be cut. Following the cut, the pipe will be threaded and installed.

When using the center-to-center method, measurements are taken from the center of one fitting

RAISED FACE

FLAT FACE

85205-13_F17.EPS

Figure 17 Raised- and flat-faced flanges.

to the center of the next fitting. The actual length of pipe between the fittings is determined by first subtracting the center-to-face dimension, or takeout, of each fitting from the overall length. Then the thread makeup for each fitting is added back to the length to obtain the actual cut length. *Figure 20* shows these dimensions for a forged steel 2,000-pound 90-degree elbow. For this example, it is assumed that the same type of elbow will be used on each end of the pipe. The takeout dimension of the ell is 1½". The pipe size is 1-inch Schedule 80.

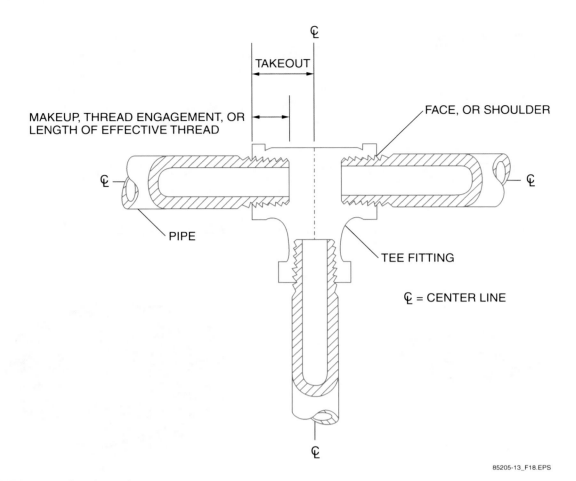

Figure 18 Takeout and makeup dimensions.

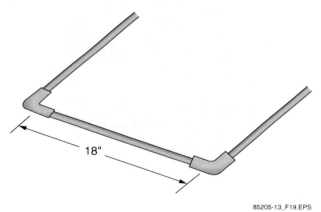

Figure 19 Center-to-center pipe measurement.

The thread engagement varies with the nominal size of the pipe. Thread engagement is best measured for accuracy. One way this can be done is by properly tightening a piece of pipe into a fitting; marking the pipe at the face of the fittings; and then removing and measuring the makeup. However, manufacturers of fittings do give dimensions of screwed fittings in makeup charts. *Table 2* provides an example of such a chart. The makeup is shown here in the final column as the

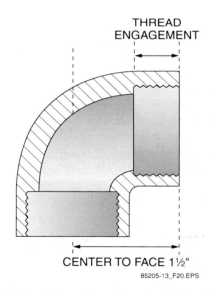

Figure 20 Center-to-face dimension of a 2,000-pound 90-degree ell.

length of effective thread. Note that the number of expected hand turns, and the length of the pipe threaded in by hand, is also listed.

Figure 21 shows the center-to-center method for determining the length of a pipe between two fit-

---

Table 2 Example of Makeup Chart

Nominal Pipe Size (Inches)	Threads per Inch	Number of Usable Threads*	Hand-Tight Engagement** (Inches)	Thread Makeup Wrench and Hand Engagement** (Inches)	Total Length of External Threads (Inches)
1/8	27	7	3/16	1/4	3/8
1/4	18	7	1/4	3/8	9/16
3/8	18	7	1/4	3/8	5/8
1/2	14	7	5/16	1/2	3/4
3/4	14	8	5/16	9/16	13/16
1	11 1/2	8	3/8	11/16	1
1 1/4	11 1/2	8	7/16	11/16	1
1 1/2	11 1/2	8	7/16	3/4	1
2	11 1/2	9	7/16	1 3/4	1 1/16
2 1/2	8	9	11/16	1 1/8	1 9/16
3	8	10	3/4	13/16	1 5/8
3 1/2	8	10	13/16	1 1/4	1 11/16
4	8	10	13/16	1 5/16	1 3/4
5	8	11	15/16	1 3/8	1 13/16
6	8	12	15/16	1 1/2	1 15/16
8	8	14	11/16	1 11/16	2 1/8
10	8	15	13/16	1 15/16	2 3/8
12	8	17	1 3/8	2 1/8	2 9/16

*Rounded to the nearest thread.
**Rounded to the nearest 1/16 inch.

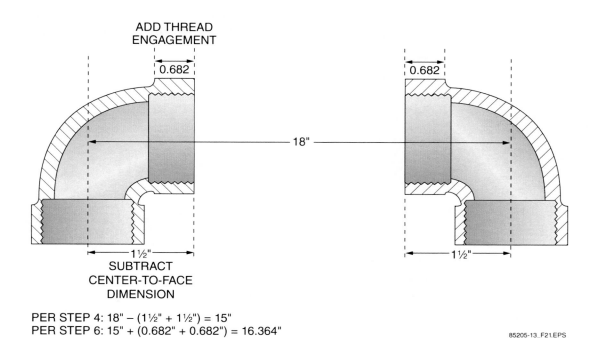

PER STEP 4: 18" − (1 1/2" + 1 1/2") = 15"
PER STEP 6: 15" + (0.682" + 0.682") = 16.364"

85205-13_F21.EPS

Figure 21 Center-to-center method of determining pipe length.

tings. Follow these steps to determine the length of a pipe between two fittings. Again, in this example, the pipe is 1-inch nominal size, the 90-degree ells are 2,000-pound rated forged steel, and the center-to-center distance is 18 inches.

Step 1 Find the center of each of the two fittings.

Step 2 Measure the distance between the centers of the two fittings (18 inches in this example).

Step 3 Measure the fitting to determine the center-to-face length, or takeout, of each of the fittings, or use a manufacturer's chart. In this example, the takeout of the fitting is 1½".

Step 4 Subtract the takeout of both fittings from the center-to-center measurement taken in Step 2.

Step 5 Determine the thread engagement of the fittings (0.682 inches in this example, per the makeup chart in *Table 2*).

> **NOTE**
>
> This can be done by using the manufacturer's makeup chart or by actually measuring the thread engagement.

Step 6 Add the thread engagement of both fittings back to the length found in to find the length of pipe needed. The final cut length of the pipe would be 16.364 inches.

This measurement is just 0.011 ($^{11}/_{1000}$ths) inches short of 16⅜".

4.2.0 Center-to-Face Method

The center-to-face method is used when the drawing gives the length from the center of a fitting, such as a 90-degree ell or a tee, to the face of a flange. To find the length of pipe needed, subtract the takeout, or laying lengths, of the fitting and the flange.

Refer to *Figure 22*. Follow these steps to determine the length of pipe between two fittings, using the center-to-face method. In this example, the distance between the center of the ell and the raised face of the flange will be 18 inches. The flange is 1" raised-face Class 300 forged steel. Assume that the pipe remains a 1-inch nominal size, and the ell remains a 1-inch 2,000-pound forged steel type.

Step 1 Find the center of the fitting.

Step 2 Measure the distance from the center of the fitting to the face of the flange (18 inches in this example).

Step 3 Measure to determine the takeout of the elbow fitting, or use a manufacturer's chart. In this example, the takeout of the fitting is 1½".

Step 4 Determine the thread engagement of the elbow fitting. Per the makeup chart, the thread engagement is 0.682".

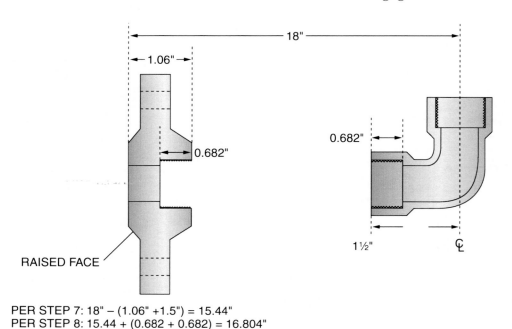

PER STEP 7: 18" − (1.06" + 1.5") = 15.44"
PER STEP 8: 15.44 + (0.682 + 0.682) = 16.804"

85205-13_F22.EPS

Figure 22 Center-to-face method of determining pipe length.

Step 5 Measure the flange to determine the total thickness or use a manufacturer's chart. The thickness of this flange is 1.06". This dimension can also be considered the takeout dimension of a flange.

Step 6 Determine the thread engagement for the flange. Since the flange thread details are the same as that of the 90-degree elbow, the thread engagement should also be 0.682" for the flange.

Step 7 Subtract the thickness of the flange and the center-to-face measurement of the elbow from the measurement taken in Step 2.

Step 8 Add back the required thread engagement (makeup) for each of the two fittings to the length found in step 7. The final cut length of the pipe would be 16.804". This is just 0.008 ($^8/_{1000}$ths) short of $16^{13}/_{16}$".

4.3.0 Face-to-Face Method

The face-to-face method is used to determine the length of a pipe that has a flange on each end. The measurement is taken from the face of one flange to the face of the other flange. This method duplicates part of the center-to-face method.

Follow these steps to determine the length of pipe between two flanges, using the face-to-face method:

Step 1 Measure the distance between the faces of the two flanges.

Step 2 Measure the flanges to determine the total thickness, or takeout, of each one or use a manufacturer's chart. The two flanges may be the same thickness, but they also may be different.

Step 3 Determine the thread engagement for the two flanges. Again, they may be the same, but they may also be different.

Step 4 Subtract the thickness of both flanges from the total measurement taken in Step 1.

Step 5 Add back the thread engagement required for both flanges to the value determined in Step 4. The result will be the cut length for the pipe.

The face-to-face method of measurement can also refer to measuring between the actual face of two fittings, such as two elbows or an elbow and a tee (*Figure 23*). In this case, determining the cut length of the pipe requires that the thread engagement for each end be added to the face-to-face measurement.

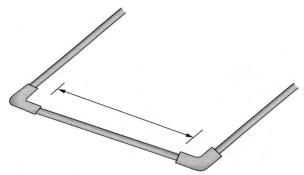

FACE-TO-FACE MEASUREMENT +
THREAD ENGAGMENT = CUT LENGTH

85205-13_F23.EPS

Figure 23 Face-to-face measurement between two fittings.

5.0.0 THREADED PIPE ASSEMBLY

Assembly techniques include selecting the proper joint compound, selecting the proper tools, and assembling various pipe sections, fittings, and components together.

5.1.0 Pipe Joint Tape and Compounds

A good grade of pipe joint compound must be applied to the male threads to prevent **galling** of the threads, ease the friction in assembling the joint, and to aid in sealing minute passages between threads. Be careful when selecting a joint compound to avoid product contamination or potentially dangerous combinations of materials. The type of joint compound to be used is often specified in project documents. Depending on the type of liquid being transferred through the pipes, sometimes no compound is permitted. Always check the specifications before applying any type of pipe joint compound. PTFE tape is often used. Various other pipe joint compounds are also used.

5.1.1 PTFE Tape

PTFE tape is a special tape that is wrapped around threads to minimize friction and help create a leak-proof joint. PTFE is an acronym for polytetrafluoroethylene. Teflon® is one brand name for PTFE.

PTFE tape (*Figure 24*) is available in several different colors to designate different uses. White PTFE is a single-density tape (about 0.7 grams/cubic centimeter), while yellow is double-density and thicker. Pink PTFE is also double-density, but measures no thicker than the white tape. Yellow PTFE is often specified for use with natural gas and other fuel gases.

When applying PTFE tape, as well as pipe joint compounds, it is essential to ensure that the material is not applied to the first thread. In addition,

85205-13_F24.EPS

Figure 24 White, yellow, and pink PTFE tape.

no material should be allowed to extend beyond the threads. Refer to *Figure 25*. Bits of PTFE tape or pipe joint compound cannot be allowed to enter the piping system.

As a general rule, use ½-inch wide PTFE tape for pipe that is ¾-inch nominal size and smaller. Use ¾-inch wide tape for pipe that is 1-inch nominal size or larger. Follow these steps to properly apply PTFE tape:

Step 1 Remove all excess cutting oil from the threads to improve the grip of the tape on the threads.

Step 2 Start the tape from the end of the thread, leaving the first full thread bare to prevent the tape from overhanging the end of the pipe.

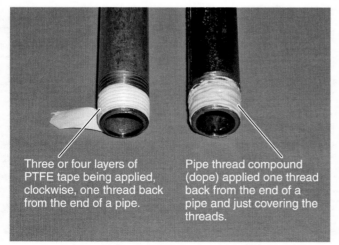

Three or four layers of PTFE tape being applied, clockwise, one thread back from the end of a pipe.

Pipe thread compound (dope) applied one thread back from the end of a pipe and just covering the threads.

85205-13_F25.EPS

Figure 25 PTFE tape and pipe joint compound applied.

Step 3 Wrap the tape around the pipe in the direction that the joint is to be assembled. Viewing the pipe from the thread end, the tape should be wrapped around right-hand threads in a clockwise direction (by far the most common) and around left-hand threads in a counterclockwise direction.

Step 4 Continue to wrap the tape around the joint, overlapping the edges of each wrap about 30 to 50 percent until all remaining effective threads have been covered.

Step 5 Press the tape against the threads to seal it to the threads and prevent it from simply slipping off the threads once joint make up has started.

5.1.2 Pipe Joint Compounds

Pipe joint compounds (*Figure 26*), also known as pipe dope, are manufactured for lubricating and sealing threaded pipe joints. It is very important to ensure that the product is suitable for both the pipe material and the fluid the piping will carry. Manufacturers provide specific data regarding the application of pipe joint compounds to ensure that users make the best possible choice. Pipefitting veterans often have a specific product that they prefer and in which they have a great deal of confidence. However, in many cases, the joint compound that must be used will be part of the project specifications.

There are many different types of pipe joint compounds on the market. Common ingredients include various types of clay and calcium carbonate. Each product typically has a submittal sheet associated with it that specifies the pipe material it can be used on, the fluids and chemicals that it works well with, and the major ingredients. Properties such as the pressure and temperature rating and any precautions that apply are also shown. A

85205-13_F26.EPS

Figure 26 Pipe joint compounds.

number of pipe joint compounds are made with PTFE as a major ingredient.

Pipe joint compound is relatively easy to apply. It is simply brushed onto the male threads to be fitted. It is best to use a brush that is stiff enough to push the compound down to the root of the threads, rather than simply brushed across the upper surface. Do not apply the compound to the female threads, and do not apply it to the first thread of the pipe end. This is to keep the material out of the pipeline and avoid contaminating the liquid inside.

5.2.0 Selecting Wrenches

When you assemble threaded pipe joints, first assemble the fitting by hand, and then make up the joint completely using a pipe wrench. Either the pipe should be secured in a vise, or two pipe wrenches should be used: one to hold the pipe and the other to turn the fitting.

Selecting the correct wrench size is important to ensure that the joint is properly tightened. A wrench that is too small does not provide sufficient leverage. A wrench that is too large allows even a small, lightweight worker to overtighten a joint with ease. Refer to *Table 3* to assist in selecting the proper wrench for a given pipe size. Using an extension on the wrench handle, such as a length of pipe often referred to as a cheater bar, is not allowed. The extension can slip off or the handle of the wrench can fail, causing an injury. Never use a pipe wrench that has a bent handle. Wrenches that are bent should be immediately taken out of service. Pipe wrenches that have worn teeth or other moving parts can often be repaired rather than replaced, especially if it is a large size.

Table 3 Suggested Wrench-to-Pipe Sizes

Wrench Size	Pipe Nominal Size
6"	⅛" – ½"
8"	¼" – ¾"
10"	¼" – 1"
12"	½" – 1½"
14"	½" – 1½"
18"	1" – 2"
24"	1½" – 2½"
36"	2" – 3½"
48"	3" – 5"
60"	3" – 8"

CAUTION

Never use cheater bars (pipes used to extend wrenches) because they can cause equipment damage. Always use the right-sized wrench to avoid damaging the pipe and the wrench.

In selecting and using the pipe wrench, use the following suggestions as a guide:

- Pipe wrenches should be in good working condition and adjusted properly to avoid damaging the pipe.
- Use a strap wrench when working with brass, chrome-plated, or other finished pipe to avoid scratching or marring the surface.
- Never use a power drive or threading machine to tighten fittings onto pipes.
- Never use a cheater bar on the end of a wrench or strike the wrench handle with a hammer.

5.3.0 Fitting Screwed Pipe and Fittings

Proper fitting of pipelines is a skill that must be learned through experience. It is crucial to fabricating a quality piping system. Poor fitting practices can result in leaks and other defects in the system. When tightening fittings onto a piece of pipe, the pipefitter must be aware of the direction that the fitting must face when the joint is complete. If the fitting is turned past the desired end position and then turned back, the fitting will probably leak.

The joining, or makeup, of a taper-threaded pipe connection is performed in two distinct operations, known as hand engagement and wrench makeup. Hand engagement on properly cut threads is normally between four and five complete revolutions. Hand engagement is a good way to check the quality of threads after cutting. Wrench makeup is usually about three turns for a total of approximately seven to eight rotations for complete makeup. More turns are required for larger pipe.

Follow these basic steps to fit screwed pipe and fittings:

Step 1 Determine the type, size, and schedule of pipe being used in the system.

Step 2 Determine the length of pipe needed.

Step 3 Cut the pipe to the desired length.

Step 4 Ream the pipe to remove all internal burrs.

Step 5 Thread the pipe, taking precautions not to cut the threads too deep.

Step 6 Select the proper size, shape, and type of fittings needed.

Step 7 Clean the pipe and fittings thoroughly, inside and out. All sand, dirt, and oil must be removed from the inside of the pipe and fittings to avoid contaminating the system or clogging the line. The fittings and pipe ends may be cleaned with a clean rag soaked in nonflammable solvent.

Step 8 Check the threads on the pipe and the fitting to ensure that they are properly cut and not damaged.

Step 9 Apply joint compound or PTFE tape to the pipe threads. Follow the project specifications.

Step 10 Start the fitting on the pipe by hand and tighten it as far as possible without significant effort.

Step 11 Tighten the fitting slowly, using a pipe wrench, ensuring that it is facing in the proper direction when tight.

5.4.0 Installing Threaded Valves

The threaded joint is a common method for joining pipe to smaller-sized valves. In order to have a strong, leak-proof threaded joint, the threads must be clean and smoothly cut. They must have the correct pitch, lead, taper, and form, and the threads must be correctly sized. If these conditions are not met, the seal will probably leak. In addition, threads that do not meet certain levels of quality are rejected in most cases.

Special precautions must be taken when installing valves with threaded ends to avoid damaging the valve. The steps are basically the same as those for installing threaded fittings. Always consult the manufacturer's guidelines for proper installation before proceeding. Follow these guidelines when installing valves with threaded ends:

- Never place the valve in any type of vise. Always secure the pipe in a vise, and screw the valve onto the pipe.
- Always grip the valve with the pipe wrench on the flats provided for the wrench, if the valve was constructed this way. Some valves have no wrench flats. In this case, place the wrench on the valve body. Do not try to use the leverage of the valve handle to turn or tighten the valve.
- Use a strap wrench on valves with brass or polished bodies to avoid excessive marring of the valve.
- Apply pipe joint compound only to the male threads of the pipe. Do not apply to the female threads inside the valve, and do not apply to the thread nearest the end of the pipe.
- Do not run the male end of the pipe all the way into the valve. This will damage the seat and cause the valve to leak. Valves may be made of a material that is significantly softer than the steel pipe. Excessive effort or using a pipe wrench that is too large can easily result in threading the pipe too far into the valve. Always leave three threads showing outside the valve. *Figure 27* shows proper thread makeup.

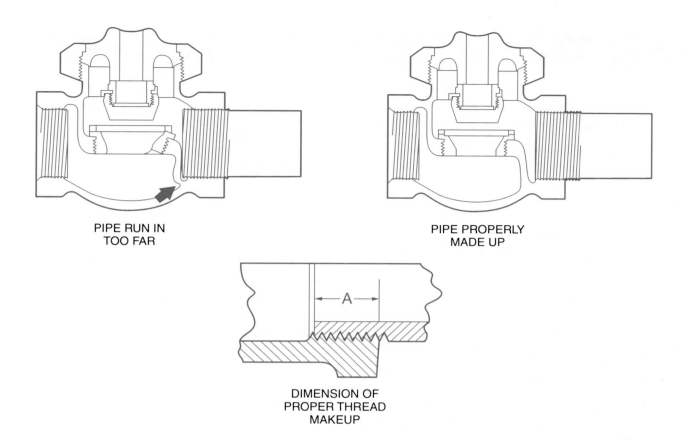

DIMENSIONS, IN INCHES									
DIMENSIONS GIVEN DO NOT ALLOW FOR VARIATIONS IN TAPPING OR THREADING									
PIPE SIZE	⅛	¼	⅜	½	¾	1	1¼	1½	2	2½
A	¼	⅛	⅜	½	9⁄16	11⁄16	11⁄16	11⁄16	¾	15⁄16

Figure 27 Proper thread makeup for a valve.

SUMMARY

Threaded pipe is often used in the maritime environment to connect smaller piping systems and components. All pipe must be threaded according to specific standards. The tapered pipe thread, or NPT, is the most common thread used on pipe and fittings for maritime applications. A threaded piping system typically requires a joint compound and proper assembly to ensure a tight, leak-proof joint.

Much of the cutting, reaming, and threading of screwed pipe systems is done in the field by pipefitters working from piping drawings. The ability to take accurate field measurements between fittings and components is an essential skill. In addition, maritime pipefitters must be able to make accurate calculations using those measurements to ensure that pipe sections are cut to the proper length. Once the pipe has been cut, it must be properly threaded and installed. There is little room for error in every task performed by the maritime pipefitter.

1. The most common thread standard used in the United States to seal against pipe leakage is _____.
   a. NPT
   b. NPS
   c. BSPP
   d. BSPT

2. In the thread designation ⅜–18 NPT, the 18 represents the number of threads per inch.
   a. True
   b. False

3. The acronym MIP stands for _____.
   a. Manufactured Iron Pipe
   b. Male Internal Pipe
   c. Male Internal Pitch
   d. Male Iron Pipe

4. The type of threaded fitting that is normally required for maritime applications is made from _____.
   a. malleable iron
   b. cast iron
   c. forged steel
   d. cast steel

85205-13_RQ01.EPS
Figure 1

5. The fitting shown in *Review Question Figure 1* is a _____.
   a. cap
   b. union
   c. coupling
   d. bushing

6. The fitting that is used to connect the male end of a pipe to the female end of a larger fitting is called a _____.
   a. reducing coupling
   b. raised-face flange
   c. reducing bushing
   d. thread insert

7. The term *takeout* refers to the _____.
   a. amount of pipe that screws into a threaded fitting
   b. distance from the shoulder of a fitting and the center of the fitting
   c. distance between the threaded ends of the pipes connected to the straight-through run of the tee
   d. shoulder-to-shoulder dimension of the fitting

8. The method used to determine the cut length for a pipe that has a flange on each end is the _____.
   a. shoulder-to-shoulder method
   b. face-to-face method
   c. center-to-center method
   d. center-to-face method

9. PTFE tape should always be used on piping that will carry steam.
   a. True
   b. False

10. Given properly formed threads, a pipefitter should be able to achieve complete makeup in about _____.
    a. 1 to 2 rotations
    b. 2 to 3 rotations
    c. 4 to 5 rotations
    d. 7 to 8 rotations

Trade Terms Introduced in This Module

Banded fitting: A pipe fitting that has a raised shoulder, or band, formed around the threaded opening to provide additional strength. Forged steel fittings are not banded.

Cast iron: A brittle, hard alloy of carbon and iron.

Ductile: A characteristic of metal that allows it to be fashioned into another form or drawn out.

Forged steel: Steel that is formed by heating it to a high temperature, but without reaching the liquid state. While in a plastic-like state, the hot steel is then pressed, hammered, or rolled into the desired shape.

Galling: Deformity of the threads in which some of the thread material of one component is removed and transferred to the threads of the other, destroying their integrity.

Length of effective thread: The portion of a threaded pipe that intrudes into a fitting or threaded component; also known as *makeup* or *thread engagement*.

Makeup: The portion of a threaded pipe that intrudes into a fitting or threaded component; also known as the *length of effective thread* or *thread engagement*.

Malleable iron: Metallic fitting material that typically has a pressure rating of 125 to 150 psi when used for fittings. Due to its lack of strength and brittle nature, it is rarely used in maritime applications.

National Pipe Thread (NPT): The US standard for pipe threads. The other primary US standard, National Pipe Straight (NPS), has no ability to seal and is used in simple structural and assembly applications only.

Takeout: Refers to the distance from the center of a fitting, such as an elbow, to the face or shoulder of the threaded opening.

Thread angle: The angle formed between two inclined faces of a thread.

Thread crest: The top of a fabricated thread. The thread crest of an NPT thread, although it may appear sharp, actually has a very slight flattening at the crest.

Thread engagement: The portion of a threaded pipe that intrudes into a fitting or threaded component; also known as *makeup* or *length of effective thread*.

Additional Resources

This module presents thorough resources for task training. The following resource material is suggested further study.

Audel Mechanical Trades Pocket Manual. Thomas B. Davis, Carl A. Nelson. New York, NY: Macmillan & Company.

IPT's Pipe Trades Handbook. Robert A. Lee. IPT Publishing and Training Ltd.

Pipefitters Handbook. Forrest R. Lindsey. Industrial Press, Inc.

Figure Credits

NCCER CURRICULA — USER UPDATE

NCCER makes every effort to keep its textbooks up-to-date and free of technical errors. We appreciate your help in this process. If you find an error, a typographical mistake, or an inaccuracy in NCCER's curricula, please fill out this form (or a photocopy), or complete the online form at **www.nccer.org/olf**. Be sure to include the exact module ID number, page number, a detailed description, and your recommended correction. Your input will be brought to the attention of the Authoring Team. Thank you for your assistance.

Instructors – If you have an idea for improving this textbook, or have found that additional materials were necessary to teach this module effectively, please let us know so that we may present your suggestions to the Authoring Team.

NCCER Product Development and Revision

13614 Progress Blvd., Alachua, FL 32615

Email: curriculum@nccer.org
Online: www.nccer.org/olf

❑ Trainee Guide   ❑ Lesson Plans   ❑ Exam   ❑ PowerPoints   Other _____

Craft / Level: _____ Copyright Date: _____

Module ID Number / Title: _____

Section Number(s): _____

Description: _____

_____

_____

_____

_____

Recommended Correction: _____

_____

_____

_____

_____

Your Name: _____

Address: _____

_____

Email: _____ Phone: _____

85206-13

Fiberglass and Plastic Pipe

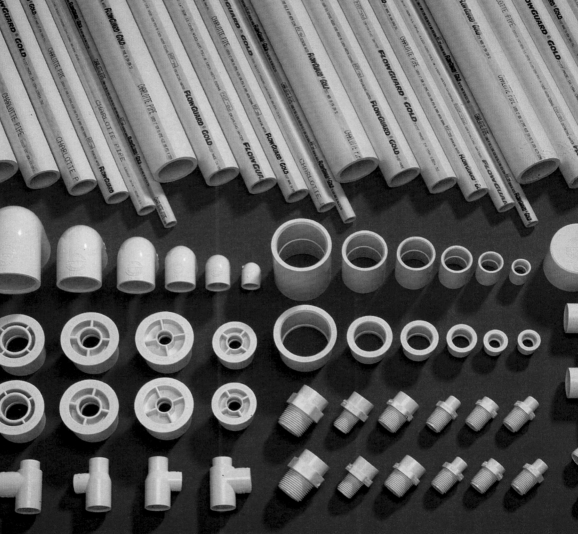

Module Six

Trainees with successful module completions may be eligible for credentialing through NCCER's National Registry. To learn more, go to **www.nccer.org** or contact us at **1.888.622.3720**. Our website has information on the latest product releases and training, as well as online versions of our *Cornerstone* magazine and Pearson's product catalog.

Your feedback is welcome. You may email your comments to **curriculum@nccer.org**, send general comments and inquiries to **info@nccer.org**, or fill in the User Update form at the back of this module.

This information is general in nature and intended for training purposes only. Actual performance of activities described in this manual requires compliance with all applicable operating, service, maintenance, and safety procedures under the direction of qualified personnel. References in this manual to patented or proprietary devices do not constitute a recommendation of their use.

Objectives

When you have completed this module, you will be able to do the following:

1. Identify the types, sizes, and assembly methods for fiberglass pipe and fittings.
2. Identify the types, sizes, and assembly methods for plastic pipe and fittings.

Performance Task

Under the supervision of the instructor, you should be able to do the following:

1. Properly measure, cut, and join plastic piping.

Trade Terms

Bell-and-spigot
Cellular core wall
Chamfer
Chemically inert
Hydrostatic pressure test
Interference fit
Pressure rating

Primer
Solid wall
Solvent weld
Thermoplastic
Witness mark
Working life

Industry-Recognized Credentials

If you are training through an NCCER-accredited sponsor, you may be eligible for credentials from NCCER's Registry. The ID number for this module is 85206-13. Note that this module may have been used in other NCCER curricula and may apply to other level completions. Contact NCCER's Registry at 888.622.3720 or go to **www.nccer.org** for more information.

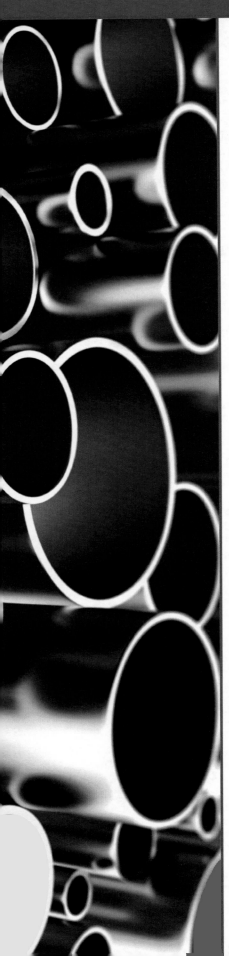

Contents

Topics to be presented in this module include:

1.0.0 Introduction...1
2.0.0 Fiberglass Pipe...1
   2.1.0 Adhesives...2
   2.2.0 Assembling Bell-and-Spigot Joints.....................................2
3.0.0 Plastic Pipe...5
   3.1.0 Plastic Pipe Sizes...5
   3.2.0 Labeling (Markings) 5
   3.3.0 Fittings..5
   3.4.0 PVC Pipe..6
   3.5.0 CPVC Pipe..7
   3.6.0 Joining Plastic Pipe..8
      3.6.1 Solvent-Cement Products...9
      3.6.2 Solvent-Cementing Plastic Pipe 10
   3.7.0 Plastic Pipe Support Spacing...12
4.0.0 Pressure Testing 14

Figures and Tables

Figure 1   Examples of fiberglass pipe applications...2
Figure 2   Bondstrand® fittings 2
Figure 3   Bondstrand® pipe joining methods...3
Figure 4   A length of Bondstrand® pipe showing spigot ends 3
Figure 5   Examples of pipe shavers...4
Figure 6   Driving pipes together 4
Figure 7   Typical pipe marking 6
Figure 8   Fittings used with PVC and CPVC pipe 7
Figure 9   PVC pipe and fittings...7
Figure 10 CPVC pipe and fittings 8
Figure 11 Plastic tubing and pipe cutters 9
Figure 12 Pipe wraparound 9
Figure 13 Deburring tools 10
Figure 14 PVC/CVPC clear pipe cleaner 10
Figure 15 PVC/CVPC primer/cleaner with a purple color 10
Figure 16 A variety of PVC and CPVC cement products 11
Figure 17 Fabricating a solvent-cemented joint 13

Table 1A   Dimensions of PVC and CPVC Schedules 40 and 80 Pipe..............6
Table 1B   Dimensions of PVC and CPVC Schedules 40 and 80 Pipe..............6
Table 2   Average Handling and Set-Up Times for PVC/CVPC Cements.......11
Table 3   Average Cure Times For PVC/CVPC Cements.............................12
Table 4   General Guidelines for Horizontal
             Support Spacing—Plastic Pipe...14

1.0.0 INTRODUCTION

Fiberglass and plastic pipe have many uses in the maritime industry. Both fiberglass and plastic pipe are lighter than carbon steel pipe, so they make an important contribution to reducing the overall weight of a vessel. They are also less expensive than metal pipe and easier to assemble because they do not require the services of certified welders. Although there are several types of plastic pipe, including ABS, PE, and HDPE, the types commonly used in the maritime industry are PVC and CPVC. The advantage of CPVC is that it is designed to withstand higher temperatures than PVC.

Plastic pipe is strong and durable, and requires little maintenance. Although it is easier to use than metal pipe, it does have some disadvantages. It can be affected by temperatures, give off toxic fumes in a fire, and become flammable. Vacuum-type sewage systems may be made from plastic pipe with adhesive joints. PVC and CPVC may also be used in potable water and water-making systems.

> **NOTE**
>
> Keep in mind that pipefitters are not permitted to make judgment calls regarding the type of pipe to use in a given situation. Just because PVC pipe is used for potable (drinkable) water in one location, for example, does not mean it can be used for that purpose in another. The project specification always governs. If there is any doubt, ask a supervisor, rather than jumping to a conclusion.

2.0.0 FIBERGLASS PIPE

Fiberglass pipe is often used for ballast and brine systems, as well as seawater cooling systems. It is typically used in sizes ranging from 2 to 14 inches, although it is available in sizes up to 40 inches. Fiberglass pipe is made from an epoxy resin reinforced with strands of fiberglass. It is commonly known as glass reinforced epoxy, or GRE. Among its benefits is that it is able to flex under mechanical stress without causing structural damage to the pipe. It is also less expensive than copper-nickel pipe, easier to install, and lighter in weight than metal pipe.

Bondstrand® is a brand of GRE pipe made by NOV Fiber Glass Systems. It is widely used on ships and offshore rigs. *Figure 1* shows examples of installations using this product. Bondstrand® pipe is used in a number of applications on ships and offshore rigs, including the following:

- Ballast water
- Brine transmission
- Chemical processing
- Cooling water
- Corrosive liquid transmission
- District heating
- Food processing
- General water service
- Potable water
- Process water
- Salt water disposal
- Seawater supply
- Sprinkler systems
- Crude oil washing
- Steam condensate
- Waste (gray) water
- Water transmission
- Water treatment
- Engine cooling

A significant advantage of GRE pipe over carbon steel pipe is its resistance to corrosion. For that reason, it is commonly used in systems that handle seawater. *Figure 2* shows fittings used with Bondstrand® pipe:

- Standard and reducing tees
- 30-, 45-, 60-, and 90-degree elbows
- Concentric reducers
- Nipples
- Flanges

A standard selection of valves is also available. Bondstrand® GRE pipe used in maritime pipefitting work is assembled using one of five methods (*Figure 3*):

- **Bell-and-spigot,** also known as Quick-Lock (*Figure 3A*), has a hub on one end and a smooth surface on the other. It is used with pipe in the 2- to 16-inch range. *Figure 4* shows an installation in process. In the background is a length of pipe on which both ends have been shaved in preparation for installation (see arrow).
- Taper-taper (*Figure 3B*) is used with pipe in the 2- to 40-inch range.
- Double O-ring (*Figure 3C*) is used when a mechanical connection is required.
- Double O-ring expansion coupling (*Figure 3D*) is used in piping systems subject to expansion.
- Flange connection (*Figure 3E*).

In addition to these methods, some GRE pipe uses a threaded connection.

All methods require the use of a specialized adhesive to bond the pipe section together or, in the case of the flange connection, to bond the flange to the pipe. Some connections that use O-rings also use a nylon dowel to lock the joint. The

HEAVY-BODIED CLEAR
PVC CEMENT

HEAVY-BODIED, FAST-DRYING
GRAY PVC CEMENT

HEAVY-BODIED ORANGE
CPVC CEMENT

85206-13_F16.EPS

Figure 16 A variety of PVC and CPVC cement products.

Use special care when solvent-cementing plastic pipe in low temperatures (below 40°F/4°C) or extremely high temperatures (above 100°F/38°C). In extremely hot environments, make sure both surfaces to be joined are still wet with cement when they are put together. Cements generally set up faster in higher temperatures. Longer curing times are likely to be required in cold temperatures. *Table 2* provides some general guidelines for handling and setup times at different temperatures, based on the pipe size. Once the joint has been made, the cement must be allowed to cure before it can withstand pressure testing. *Table 3* provides the average cure times for both PVC and CPVC pipe cements. Note that the times shown in both tables can, and should be, extended 50 percent in wet or extremely humid environments. If this represents a problem, designers or supervisors will select another cement that bonds and cures faster in humid conditions.

Figure 17 shows a logical sequence of steps to fabricate a solvent-cement joint for PVC and CPVC pipe. Note that these steps illustrate typical instructions. When joining plastic pipe, always follow the cement manufacturer's instructions for the environment and materials being joined.

Before beginning, put on the appropriate PPE. Safety glasses should be worn at all times when working with pipe and solvent-cement products. In addition, rubber or latex gloves should also be worn during the joining process. Making a proper joint requires careful timing, so make sure that all the required materials are readily available and within easy reach.

Step 1 Cut the pipe to the required length (*Figure 17A*), then ream the inside edge of the pipe with a reamer or sharp knife. **Chamfer** the outside edge of the pipe with a suitable reamer, single-cut file, or sand cloth to remove rough edges (*Figure 17B*). Chamfering involves breaking the edge of the pipe. Chamfering is good practice because it provides for a more secure joint. When the socket and pipe fit tighter, it reduces the possibility of leakage from the pipe. Examine the pipe end; if cracks are noted, the pipe should be cut off at least 2 inches away from the end of the crack.

Step 2 Dry-fit the pipe and fitting (*Figure 17C*). The pipe should fit more tightly as it reaches about two-thirds of the socket

Table 2 Average Handling and Set-Up Times for PVC/CVPC Cements

Temperature During Assembly	Pipe Diameter					
	½" to 1¼"	1½" to 3"	4" to 5"	6" to 8"	10" to 16"	18"+
60°F–100°F	2 minutes	5 minutes	15 minutes	30 minutes	2 hours	4 hours
40°F–60°F	5 minutes	10 minutes	30 minutes	90 minutes	8 hours	16 hours
20°F–40°F	8 minutes	12 minutes	96 minutes	3 hours	12 hours	24 hours
0°F–20°F	10 minutes	15 minutes	2 hours	6 hours	24 hours	48 hours

Note: Handling/setup time is the time required prior to handling the joint. In damp or humid weather, allow 50 percent additional time.

Table 3 Average Cure Times For PVC/CVPC Cements

RH 60% or Less Temp. During Assembly or Cure Period	Pipe Diameter										
	½" to 1¼"		1½" to 3"		4" to 5"		6" to 8"		10" to 16"	18"+
	Up to 180 psi	180 psi+	Up to 180 psi	180 psi+	Up to 180 psi	180 psi+	Up to 180 psi	180 psi+	Up to 100 psi	Up to 100 psi
60°F–100°F	1 hr	6 hrs	2 hrs	12 hrs	6 hrs	18 hrs	8 hrs	24 hrs	24 hrs	36 hrs
40°F–60°F	2 hrs	12 hrs	4 hrs	24 hrs	12 hrs	36 hrs	16 hrs	48 hrs	48 hrs	72 hrs
20°F–40°F	6 hrs	36 hrs	12 hrs	72 hrs	36 hrs	4 days	3 days	9 days	8 days	12 days
0°F–20°F	8 hrs	48 hrs	16 hrs	96 hrs	48 hrs	8 days	4 days	12 days	10 days	14 days

Note: Joint cure time is the time required before pressure testing the system. In damp or humid weather, allow 50 percent additional cure time.

depth. This is known as an **interference fit**. Once the primer and cement are applied, the surfaces will be mildly dissolved and slippery, and the pipe and fitting should fit together fully.

Step 3 Mark the pipe and fitting with a felt-tipped pen or scratch awl to show the proper position for alignment, if necessary. Then mark the depth of the fitting socket on the pipe. Since the mark will be removed by cleaners and primers, make what is known as a **witness mark** another 2 inches away. Once the fitting is assembled, a measurement can quickly be made from the witness mark to the face of the fitting to make sure that the pipe bottomed-out in the socket. During the first few minutes of curing time, the pipe is sometimes pushed out of the socket somewhat by natural processes. The witness mark helps to ensure that the joint stayed together during the curing time.

Step 4 Wipe off the fitting and pipe to remove any loose dirt and moisture (*Figure 17D*). Clean the surfaces being joined by applying a proper cleaning product with the provided dauber. No additional action is required other than simply applying the cleaner and working it in a bit.

Step 5 Next, apply the primer to the surfaces being joined. Work the primer in with vigor; the dauber will begin to drag some when the surface starts to break down as desired. Apply primer to the fitting socket first, and then to the end of the pipe. Apply the primer to the pipe 1 to 1½ inches beyond the socket depth. Then, apply the primer to the socket of the fitting a second time.

Step 6 Work-in a heavy, even coat of cement to the pipe end while the primer is still wet (*Figure 17E*). Apply the cement up to the depth of the socket only. Use the same applicator (without adding more cement) to apply a thin coat inside the fitting socket (*Figure 17F*). Do not allow excess cement to puddle in the fitting socket.

Step 7 Apply a second layer of cement to the pipe end.

Step 8 Immediately insert the pipe into the fitting socket and rotate the tube one-quarter to one-half turn while inserting (*Figure 17G*), assuming there is enough space to do so. This motion ensures an even distribution of cement in the joint. Properly align the fitting.

> **NOTE**
> Be careful when aligning fittings on a pipe. Ensure that all fittings are plumb.

Step 9 Hold the assembly together firmly for about 30 seconds. This helps prevent the pipe from pushing out of the socket as the curing process begins. An even bead of cement should be visible around the joint. If this bead does not appear all the way around the socket edge, the cement was improperly applied. In this case, remake the joint to avoid the possibility of leaks. Wipe excess cement from the tubing and fitting surfaces; the bead should not be wiped away.

3.7.0 Plastic Pipe Support Spacing

Since even rigid plastic piping products are somewhat flexible and are more sensitive to heat than metals, support must be provided at closer intervals. *Table 4* provides guidelines for support spac-

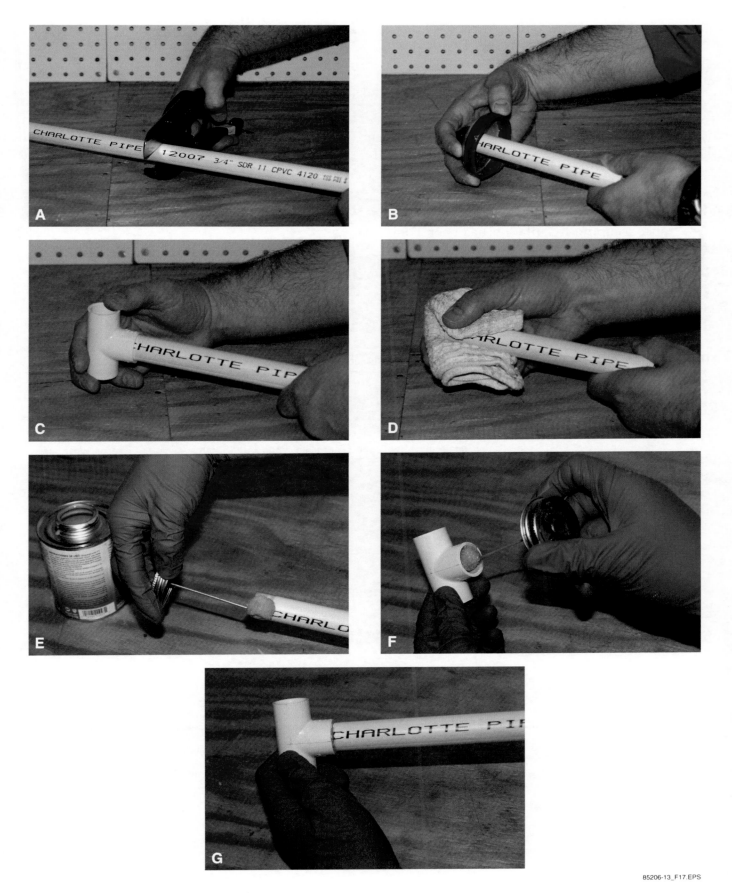

Figure 17 Fabricating a solvent-cemented joint.

85206-13_F17.EPS

ing intervals for several schedules of PVC pipe. Typical spacing is at 4-foot intervals. Smaller pipe sizes require closer intervals. It is also important to note the operating temperature of the piping system in the table. As the operating temperature increases, the support interval becomes closer. Other tables are available for higher operating temperatures for products that can safely tolerate it. Local codes may require even closer support intervals for one or more products.

4.0.0 PRESSURE TESTING

Once a piping installation has been completed and the joints have cured, a **hydrostatic pressure test** is generally conducted to verify that the system is leak-free and will operate under pressure. This process involves filling the system with water and bleeding all air out from the highest and farthest points in the run.

If a leak is found, the leaking section must be re-done. A new section can usually be installed using couplings. During subfreezing temperatures, blow water out of the lines after testing to avoid possible damage to the pipes from freezing.

CAUTION
Never pressure-test a connection until the manufacturer's recommended cure times have been met. After testing a connection, thoroughly flush the system for at least 10 minutes to remove any remaining trace amounts of solvent-cement.

WARNING!
Use air testing for plastic pipe (except PVC) only when hydrostatic testing is not practical. If air testing must be performed, never use pure oxygen.

Table 4 General Guidelines for Horizontal Support Spacing—Plastic Pipe

Nom. Pipe Size (in)	PVC Pipe															ABS Pipe				
	SDR 21 PR200 & SDR 26 PR160					Schedule 40					Schedule 80					Schedule 40				
	Operating Temp (°F)					Operating Temp (°F)					Operating Temp (°F)					Operating Temp (°F)				
	60	80	100	120	140	60	80	100	120	140	60	80	100	120	140	60	80	100	120	140
½	3½	3½	3	2		4½	4½	4	2½	2½	5	4½	4½	3	2½					
¾	4	3½	3	2		5	4½	4	2½	2½	5½	5	4½	3	2½					
1	4	4	3½	2		5½	5	4½	3	2½	6	5½	5	3½	3					
1¼	4	4	3½	2½		6	5½	5	3	3	6	6	5½	3½	3					
1½	4½	4	4	2½		6	5½	5	3½	3	6½	6	5½	3½	3½	6	6	5½	3½	3
2	4½	4	4	3		6	5½	5	3½	3	7	6½	6	4	3½	6	6	5½	3½	3
2½	5	5	4½	3		7	6½	6	4	3½	7½	7½	6½	4½	4					
3	5½	5½	4½	3		7	7	6	4	3½	8	7½	7	4½	4	7	7	7	4	3½
4	6	5½	5	3½		7½	7	6½	4½	4	9	8½	7½	5	4½	7½	7½	7	4½	4
6	6½	6½	5½	4		8½	8	7½	5	4½	10	9½	9	6	5	8½	8½	8	5	4½
8	7	6½	6	5		9	8½	8	5	4½	11	10½	9½	6½	5½					
10						10	9	8½	5½	5	12	11	10	7	6					
12						11½	10½	9½	6½	5½	13	12	10½	7½	6½					
14						12	11	10	7	6	13½	13	11	8	7					
16						12½	11½	10½	7½	6½	14	13½	11½	8½	7½					

Note:
Always follow local code requirements for hanger spacing. Most plumbing codes have the following hanger spacing requirements:
- ABS and PVC pipe have a maximum horizontal hanger spacing of every 4' for all sizes.
- CPVC pipe or tubing has a maximum horizontal hanger spacing of every 3' for 1" and under, and every 4' for sizes 1¼" and larger.

SUMMARY

Although carbon steel pipe is the most common type used in maritime applications, plastic and fiberglass pipe are used in some situations. Fiberglass pipe is commonly used in seawater systems because, unlike steel pipe, it is highly resistant to corrosion. It is also used in a variety of other systems, including fire suppression, wastewater, and freshwater. PVC and CPVC plastic pipe is likewise used in wastewater and potable water systems. Its use is limited by its lack of fire resistance. Both fiberglass and plastic pipe are assembled using adhesive, which makes their assembly process simpler than that of steel pipe, which is either welded or threaded. Fiberglass and plastic pipe are lighter than steel pipe, so handling and installation are easier.

Fiberglass pipe used in maritime applications is made in either a bell-and-spigot or taper-taper configuration. The bell and spigot design is used in the 2- to 16-inch size range, so this will be the most common method used in maritime work. When it is necessary to cut the pipes, joining of fiberglass pipe will take longer than plastic pipe because the joint has to be shaved using special equipment in order to create a spigot.

There are several types of plastic pipe and tubing on the market, but the types generally used in maritime work are PVC and CPVC. Both types are joined with adhesive and fittings, but use different adhesives. When preparing to join plastic pipe, the pipe must be cut square, deburred, and thoroughly cleaned.

1. The type(s) of plastic pipe commonly used in the maritime industry is (are) _____.
   a. ABS
   b. PE and PEX
   c. PVC and CPVC
   d. HDPE

2. The typical curing time for a fiberglass pipe adhesive joint is _____.
   a. 15 minutes
   b. 1 hour
   c. 2 to 3 hours
   d. 4 to 6 hours

3. Plastic pipe must be protected from long-term sun exposure because it will degrade over time.
   a. True
   b. False

4. Which type(s) of plastic piping can be threaded?
   a. Both PVC and CPVC
   b. Schedule 40 PVC
   c. Schedule 80 PVC
   d. Schedule 40 CPVC

5. The most common way to join plastic pipe is _____.
   a. solvent-cementing
   b. heat-bonding
   c. threading
   d. flare fitting

6. Which of the following is used to temporarily soften plastic pipe and the fitting materials before cementing?
   a. Expander tool
   b. Adapter
   c. Primer
   d. Polymer

7. Plastic pipe cements cure fastest in _____.
   a. hot environments
   b. cold environments
   c. humid environments
   d. wet environments

8. What is the average handling and setup time for a PVC/CPVC cement fitting at a temperature of 32°F (0°C) for an 8" pipe?
   a. 2 minutes
   b. 30 minutes
   c. 3 hours
   d. 48 hours

9. If any indication of damage or cracking is visible at the pipe end, cut off at least _____.
   a. 1 inch beyond any visible crack
   b. 2 inches beyond any visible crack
   c. 3 inches beyond any visible crack
   d. 4 inches beyond any visible crack

10. In general, PVC piping should be supported at a maximum interval of _____.
    a. 3 feet
    b. 4 feet
    c. 5 feet
    d. 8 feet

Trade Terms Introduced in This Module

Bell-and-spigot: Pipe that has a bell, or enlargement, also called a hub, at one end of the pipe and a spigot, or smooth end, at the other end. The bell and spigot of two different pipes slide together to form a joint.

Cellular core wall: Plastic pipe wall that is low-density, lightweight plastic containing entrained (trapped) air.

Chamfer: To break the edge of construction material.

Chemically inert: Does not react with other chemicals.

Hydrostatic pressure test: The process in which a pipe is filled with water and all air is bled out from the highest and farthest points in the run.

Interference fit: A fit that tightens as the pipe is pushed into the socket.

Pressure rating: The maximum pressure at which a component or system may be operated continuously.

Primer: A liquid applied to plastic pipe prior to solvent welding in order to pre-soften the pipe and ensure a strong solvent weld.

Solid wall: Plastic pipe wall that does not contain trapped air.

Solvent weld: A joint created by joining two pipes using solvent cement that softens the material's surface.

Thermoplastic: Pipe that can be repeatedly softened by heating and hardened by cooling. When softened, thermoplastic pipe can be molded into desired shapes.

Witness mark: A mark made for the purpose of determining the proper position of a pipe and fitting during the joining process. The witness mark is made in a position that will not be hidden when the pipe and fitting are assembled.

Working life: The time it takes for freshly mixed adhesive to begin to harden.

Additional Resources

This module presents thorough resources for task training. The following resource material is suggested for further study.

Pipefitter's Handbook, Revised Edition, 2009. Howard C. Massey. Carlsbad, CA: Craftsman Book Company.

Figure Credits

Courtesy of Charlotte Pipe and Foundry, Module opener, Figures 9, 10, 17, Table 4

Courtesy of NOV Piping Systems, Figures 1, 2, 3, 4, 5A

Topaz Publications, Inc., Figure 5B

Tables courtesy GF Harvel LLC, an entity under the Division Americas of Georg Fischer Piping Systems. ©2012 GF Harvel; reprinted with permission-All Rights Reserved, Tables 1A, 1B

LASCO Fittings, Inc., Figure 8 (all except reducer)

Reed Manufacturing Company, Figure 11A, C, D

Courtesy of RIDGID®, Figure 11B

Courtesy of Flange Wizard®, Figure 12

Courtesy of Imperial® Brand Tools - Stride Tool Inc., Figure 13

Courtesy of Oatey, Figures 14, 15, 16, Tables 2 and 3

NCCER CURRICULA — USER UPDATE

NCCER makes every effort to keep its textbooks up-to-date and free of technical errors. We appreciate your help in this process. If you find an error, a typographical mistake, or an inaccuracy in NCCER's curricula, please fill out this form (or a photocopy), or complete the online form at **www.nccer.org/olf**. Be sure to include the exact module ID number, page number, a detailed description, and your recommended correction. Your input will be brought to the attention of the Authoring Team. Thank you for your assistance.

Instructors – If you have an idea for improving this textbook, or have found that additional materials were necessary to teach this module effectively, please let us know so that we may present your suggestions to the Authoring Team.

NCCER Product Development and Revision

13614 Progress Blvd., Alachua, FL 32615

Email: curriculum@nccer.org
Online: www.nccer.org/olf

❏ Trainee Guide   ❏ Lesson Plans   ❏ Exam   ❏ PowerPoints   Other _____

Craft / Level: _____   Copyright Date: _____

Module ID Number / Title: _____

Section Number(s): _____

Description: _____

_____

_____

_____

_____

Recommended Correction: _____

_____

_____

_____

_____

Your Name: _____

Address: _____

_____

Email: _____   Phone: _____

85207-13

Identifying Valves, Flanges, and Gaskets

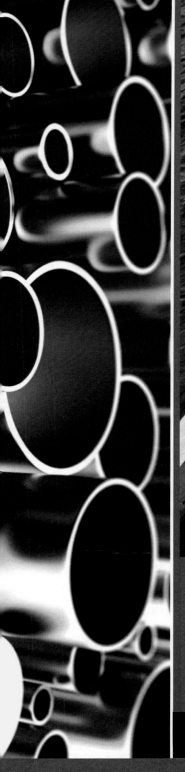

Module Seven

Objectives

When you have completed this module, you will be able to do the following:

1. Identify various types of valves based on their primary functions.
2. Identify types of valve actuators.
3. Describe factors related to the selection, storage, and handling of valves.
4. Identify and describe various types of flanges and gaskets.

Performance Tasks

This is a knowledge-based module; there are no performance tasks.

Trade Terms

Actuator	Phonographic
ASTM International	Plug
Angle valve	Plug valve
Ball valve	Positioner
Body	Pressure differential
Bonnet	Pressure vessel
Butterfly valve	Reach rod
Cavitation	Relief valve
Check valve	Seat
Control valve	Severe service
Corrosive	Thermal transients
Deformation	Throttling
Disc	Torque
Elastomeric	Trim
Galling	Turbulence
Gasket	Valve body
Gate valve	Valve stem
Globe valve	Valve trim
Head loss	Wedge
Kinetic energy	Wire drawing
Packing	Yoke bushing

Industry-Recognized Credentials

If you are training through an NCCER-accredited sponsor, you may be eligible for credentials from NCCER's Registry. The ID number for this module is 85207-13. Note that this module may have been used in other NCCER curricula and may apply to other level completions. Contact NCCER's Registry at 888.622.3720 or go to **www.nccer.org** for more information.

Contents

1.0.0 Introduction 1
2.0.0 Valves That Start and Stop Flow 1
    2.1.0 Gate Valves 1
        2.1.1 Valve Stem 5
    2.2.0 Knife Gate Valves 5
    2.3.0 Ball Valves 6
    2.4.0 Plug Valves 8
    2.5.0 Three-Way Valves 12
3.0.0 Valves That Regulate Flow 12
    3.1.0 Globe Valves 12
    3.2.0 Y-Type Valves 14
    3.3.0 Butterfly Valves 14
        3.3.1 Wafer Valves 16
        3.3.2 Wafer Lug Valves 16
        3.3.3 Two-Flange Valves 16
    3.4.0 Diaphragm Valves 16
    3.5.0 Needle Valves 17
4.0.0 Control Valves 17
5.0.0 Valves That Relieve Pressure 19
    5.1.0 Safety Valves 19
    5.2.0 Pressure-Relief Valves 19
6.0.0 Valves that Regulate Direction of Flow 19
    6.1.0 Swing Check Valves 20
    6.2.0 Lift Check Valves 21
    6.3.0 Ball Check Valves 22
    6.4.0 Butterfly Check Valves 22
    6.5.0 Foot Valves 23
7.0.0 Valve Actuators 24
    7.1.0 Gear Operators 24
        7.1.1 Spur-Gear Operators 24
        7.1.2 Bevel-Gear Operators 24
        7.1.3 Worm-Gear Operators 24
    7.2.0 Chain Operators 25
    7.3.0 Pneumatic and Hydraulic Actuators 27
    7.4.0 Electric or Air Motor-Driven Actuators 27
    7.5.0 Reach Rods 27
8.0.0 Storing and Handling Valves 28
    8.1.0 Safety Considerations 28
    8.2.0 Storing Valves 28
    8.3.0 Rigging Valves 28
9.0.0 Valve Placement 29
10.0.0 Valve Selection, Types, and Applications 30
    10.1.0 Valve Selection 30
    10.2.0 Valve Types and Applications 31

11.0.0 Valve Markings and Nameplate Information 31
  11.1.0 Rating Designation 32
  11.2.0 Trim Identification 32
  11.3.0 Size Designation 33
  11.4.0 Thread Markings 33
  11.5.0 Valve Schematic Symbols 33
12.0.0 Flanges 35
  12.1.0 Types of Flanges.. 35
    12.1.1 Weld Neck Flanges 35
    12.1.2 Slip-On Flanges 35
    12.1.3 Slip-On Reducing Flanges 35
    12.1.4 Blind Flanges 37
    12.1.5 Socket Weld Flanges 37
    12.1.6 Threaded Flanges 37
    12.1.7 Lap-Joint Flanges 37
    12.1.8 Silver-Brazed Flanges 38
  12.2.0 Flange Facings 38
    12.2.1 Raised-Face Flanges 38
    12.2.2 Flat-Face Flanges 39
    12.2.3 Ring Joint Type Flanges 39
    12.2.4 Tongue and Groove Flanges... 41
    12.2.5 Male and Female Flanges 41
13.0.0 Flange Gaskets... 41
  13.1.0 Gasket Materials 41
    13.1.1 PTFE Gaskets 42
    13.1.2 Fiberglass Gaskets 43
    13.1.3 Acrylic Fiber Gaskets 43
    13.1.4 Metal Gaskets 43
    13.1.5 Natural Rubber Gaskets 44
    13.1.6 Cork Gaskets 45
    13.1.7 Vinyl Gaskets 45
    13.1.8 Ceramic Gaskets 45
    13.1.9 Asbestos Gaskets 45
    13.1.10 EPDM 45
    13.1.11 Neoprene 45
    13.1.12 Nitrile 45
    13.1.13 Silicone 45
    13.1.14 Viton® 45
    13.1.15 Gylon® or Amerilon® 45
    13.1.16 Graphite-Impregnated Gaskets 46
    13.1.17 Soft Metal Gaskets 46
    13.1.18 Gasket Color Codes 46
  13.2.0 Types of Flange Gaskets 46
    13.2.1 Flat Ring Gaskets 46
    13.2.2 Full-Face Gaskets.. 46

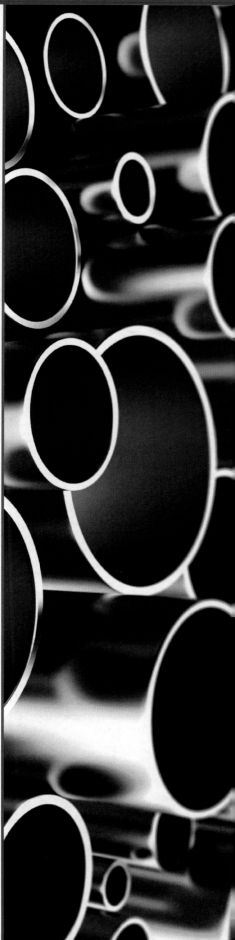

Figures and Tables

Figure 1   Gate valve 2
Figure 2   Wedge types 2
Figure 3   Solid or single wedge 3
Figure 4   Flexible wedge.. 3
Figure 5   Split wedge 3
Figure 6   Parallel-disc gate valve 4
Figure 7   Parallel disc 4
Figure 8   Spring-loaded parallel-disc gate valve (cutaway)........... 4
Figure 9   Parallel-disc gate valve with spring 5
Figure 10 Relieving disc gate valve 5
Figure 11 Valve stems 6
Figure 12 Knife gate valve 6
Figure 13 Ball valve 7
Figure 14 Ball valve with slanted seals 7
Figure 15 Venturi-type ball valve 7
Figure 16 Top-entry ball valve.. 8
Figure 17 Split-body ball valve 8
Figure 18 Plug valve 8
Figure 19 Plug configurations 9
Figure 20 Lubricated taper plug valve 10
Figure 21 Cam-operated and non-lubricated plug valve 10
Figure 22 Typical non-lubricated plug valve with elastomer sleeve............. 10
Figure 23 Three-way valve applications................................11
Figure 24 Globe valve 12
Figure 25 Angle globe valve.. 13
Figure 26 Seat/disc arrangements 13
Figure 27 Globe valve plugs.. 14
Figure 28 Y-type valve 15
Figure 29 Butterfly valve... 16
Figure 30 Wafer butterfly valve.. 16
Figure 31 Wafer lug butterfly valve...................................... 17
Figure 32 Two-flange butterfly valve 17
Figure 33 Diaphragm valve 18
Figure 34 Needle valve 19
Figure 35 Control valve.. 20
Figure 36 Control valve schematic.. 21
Figure 37 Examples of pressure-relieving valves................. 21
Figure 38 Safety valve... 22
Figure 39 Pressure-relief valve... 22
Figure 40 Swing check valves 23
Figure 41 Lift check valve.. 23
Figure 42 Ball check valve... 23
Figure 43 Butterfly check valve... 23
Figure 44 Foot valve.. 24
Figure 45 Spur-gear operator 25

2.5.0 Three-Way Valves

Three-way valves are multiport plug valves that are installed at the intersection of three lines. They are used to direct the flow between two of the lines only and to block off the third line. Situations requiring three-way valves include alternating the connections of two supply lines to a common delivery line or diverting a line into either of two possible directions. Three-way valves are designed so that when the plug is turned from one position to another, the channels previously connected are completely closed off before the new channels begin to open. This design prevents mixture of fluids or loss of pressure. Three-way valves come in several different arrangements. They can be two- or three-port valves with stops that limit the turning of the plug to two, three, or four positions (*Figure 23*).

3.0.0 VALVES THAT REGULATE FLOW

Many valves are designed to provide accurate flow control through a system. The valves used to regulate flow through a system include the following:

- Globe
- Angle globe
- Y-type
- Needle
- Butterfly
- Diaphragm
- Control

3.1.0 Globe Valves

Globe valves are used to stop, start, and regulate fluid flow. They are important elements in marine boiler systems and are commonly used as the standard against which other valve types are judged.

As shown in *Figure 24*, the globe valve disc can be totally removed from the flow path or it can completely close the flow path. The essential principle of globe valve operation is the perpendicular movement of the disc away from the seat. This causes the space between disc and seat ring to gradually close as the valve is closed. This characteristic gives the globe valve good throttling ability, which permits its use in regulating flow. Therefore, the globe valve is used not only for start-stop functions, but also for flow-regulating functions.

It is generally easier to obtain very low seat leakage with a globe valve as compared to a gate valve. This is because the disc-to-seat ring contact is close to perpendicular, which permits the force of closing to tightly seat the disc.

Globe valves can be arranged so that the disc closes against the direction of fluid flow (flow-to-open) or so that the disc closes in the same direction as fluid flow (flow-to-close). When the disc closes against the direction of flow, the **kinetic energy** of the fluid impedes closing but aids opening of the valve. When the disc closes in the same direction as flow, the kinetic energy of the fluid aids closing but impedes opening of the valve. This characteristic makes the globe valve preferable to the gate valve when quick-acting stop valves are necessary.

Along with its advantages, the globe valve has a few drawbacks. Although valve designers can eliminate any or all of the drawbacks for specific services, the corrective measures are expensive and often narrow the valve's scope of service. The most evident shortcoming of the simple globe valve is the high **head loss**, a loss of pressure from the two or more right-angle turns of flowing fluid.

High-pressure losses in the globe valve can cost thousands of dollars a year for large high-pressure lines. The fluid-dynamic effects from the pulsation, impacts, and pressure drops in traditional globe valves can damage **valve trim**, stem packing, and actuators. Troublesome noise can also result. In addition, large sizes require considerable power to operate, which may make gearing or levers necessary.

Another drawback is the large opening needed for assembly of the disc. Globe valves are often heavier than other valves of the same flow rating. The cantilever mounting of the disc on its stem is also a potential trouble source. Each of these shortcomings can be overcome, but only at costs in dollars, space, and weight.

The **angle valve** shown in *Figure 25* is a simpler modification of the basic globe form. With the ports at right angles, the diaphragm can be a simple flat plate. Fluid can flow through with only a single 90-degree turn, discharging more symmetrically than the discharge from an ordinary globe. Installation advantages also may suggest the angle valve. It can replace an elbow, for example.

For moderate conditions of pressure, temperature, and flow, the angle valve closely resembles the ordinary globe. Many manufacturers have interchangeable trim and bonnets for the two body styles, with the body differing only in the outlet end. The angle valve's discharge conditions are so favorable that many high-technology control valves use this configuration. Like straight flow-through globe valves, they are self-draining and tend to prevent solid buildup inside the valve body.

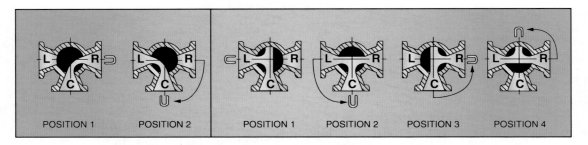

TWO-PORT, TWO POSITIONS

THREE-PORT, FOUR POSITIONS

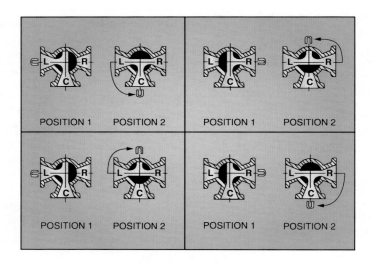

THREE-PORT, TWO POSITIONS

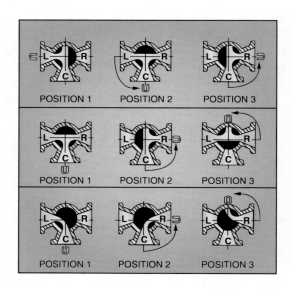
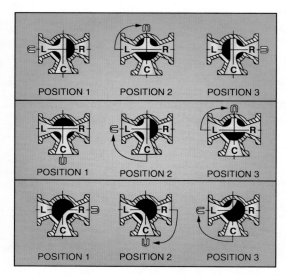

THREE-PORT, THREE POSITIONS

85207-13_F23.EPS

Figure 23 Three-way valve applications.

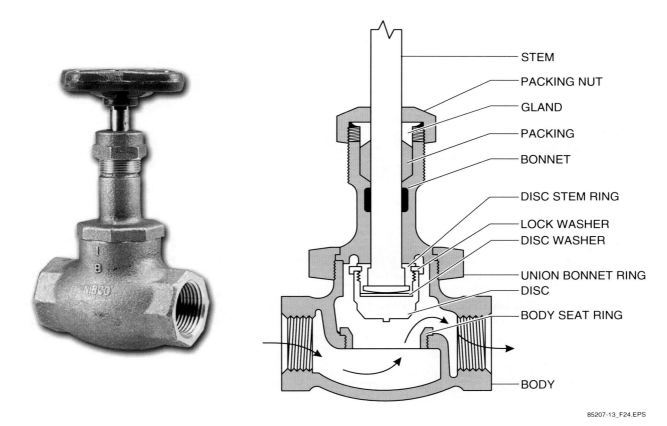

85207-13_F24.EPS

Figure 24 Globe valve.

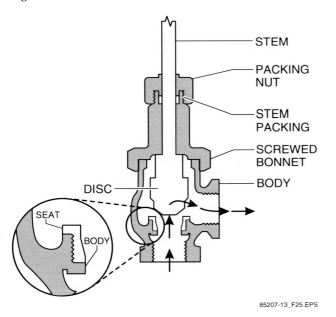

85207-13_F25.EPS

Figure 25 Angle globe valve.

Like valve bodies, there are also many variations of disc and seat arrangements for globe valves. The three basic types are shown in *Figure 26.*

The ball-shaped disc shown in *Figure 26A* fits on a tapered, flat-surfaced seat and is generally used on relatively low-pressure, low-temperature systems. It is generally used in a fully open or shut position, but it may be employed for moderate throttling of a flow.

Figure 26B shows one of the proven modifications of seat/disc design, a hard nonmetallic insert ring on the disc to make closure tighter on steam and hot water. The composition disc resists erosion and is sufficiently resilient and cut resistant to close on solid particles without serious permanent damage.

The composition disc is renewable. It is available in a variety of materials that are designed for different types of service, such as high- and low-temperature water, air, or steam. The seating surface is often formed by a rubber O-ring or washer.

The plug-type disc, *Figure 26C*, provides the best throttling service because of its configuration. It also offers maximum resistance to galling, **wire drawing**, and erosion. Plug-type discs are available in a variety of specific configurations, but in general they all have a relatively long, tapered configuration. Each of the variations has specific types of applications and certain fundamental characteristics. *Figure 27* shows the various types.

The equal percentage plug, as its name indicates, is used to allow precise control of flow through the valve. For example, if the actuator is turned 30 percent of its 90-degree swing, flow will increase by 30 percent.

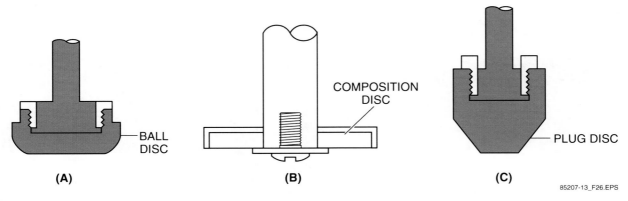

Figure 26 Seat/disc arrangements.

V-port plugs have a V-shaped port cut in the plug. The flow does not change direction, but is throttled because the port is smaller than the flow chamber. Unlike other ports, this valve allows the flow to be controlled with only moderate changes of pressure.

Needle plugs are used primarily for instrumentation applications and are seldom available in valves over 1 inch in size. These plugs lower pressure and flow. The threads on the stem are usually very fine. For that reason, the opening between the disc and seat does not change rapidly with stem rise. This permits closer regulation of flow.

All of the plug configurations are available in either a conventional globe valve design or the angle valve design. When the needle plug is used, the valve name changes to needle valve. In all other cases the valves are still referred to as globe valves with a specific type of disc.

Globe and angle valves should be installed so that the pressure is under the disc (flow-to-open). This promotes easy operation. It also helps to protect the packing and eliminates a certain amount of erosive action on the seat and disc faces. However, when high temperature steam is the medium being controlled, and the valve is closed with the pressure under the disc, the valve stem, which is now out of the fluid, contracts on cooling. This action tends to lift the disc off the seat, causing leaks that eventually result in wire drawing, small erosion channels on seat and disc faces. Therefore, in high-temperature steam service, globe valves may be installed so that the pressure is above the disc (flow-to-close).

3.2.0 Y-Type Valves

Y-type valves (*Figure 28*) are a cross between a gate valve and a globe valve. They provide the straight-through flow with minimum resistance of the gate valve and the throttling ability and flow control of a globe valve. The Y-type valve produces lower pressure drop and turbulence than a standard globe valve and is preferred over a globe valve for use in corrosive services. Y-type valves are also used in many high-pressure applications, such as boiler systems. All of the disc and seat designs available in globe valves are also available in Y-type valves.

3.3.0 Butterfly Valves

A butterfly valve has a round disc that fits tightly in its mating seat and rotates 90 degrees in one direction to open and allow fluid to pass through the valve (*Figure 29*). The butterfly valve can be operated quickly by turning the handwheel or hand lever one-quarter of a turn or 90 degrees to open or close the valve. These valves can be used

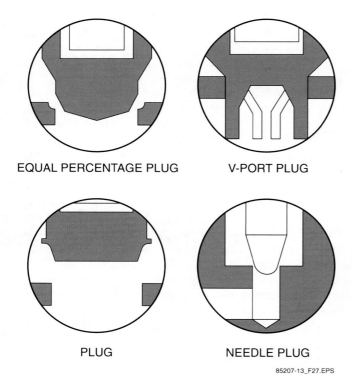

Figure 27 Globe valve plugs.

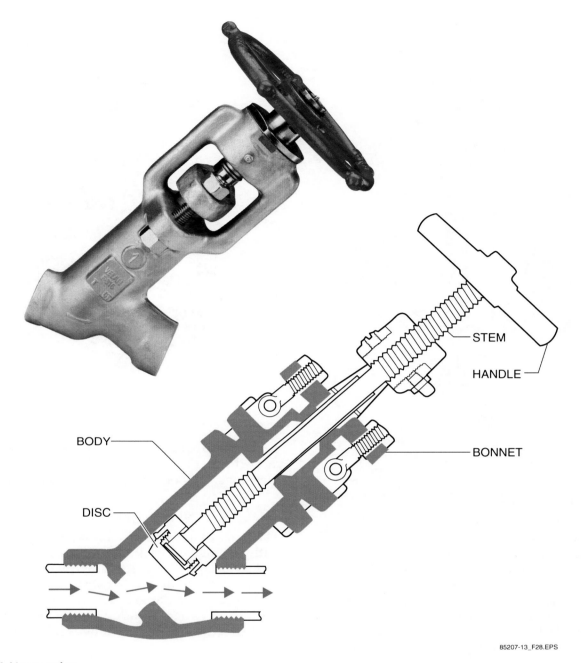

BODY

DISC

STEM

HANDLE

BONNET

85207-13_F28.EPS

Figure 28 Y-type valve.

completely open, completely closed, or partially open for noncritical throttling applications.

Butterfly valves are typically used in low-to-medium pressure and low-to-medium temperature applications. They generally weigh less than other types of valves because of their narrow body design. When the butterfly valve is equipped with a hand lever, the position of the hand lever indicates whether the valve is open, closed, or partially open. When the lever is parallel with the flow line through the valve, the valve is open. When the lever is perpendicular to the flow line through the valve, the valve is closed.

Butterfly valves that are 12 inches in diameter and larger are usually equipped with a hand-wheel or gear-operated actuator because of the large amount of fluid flowing through the valve and the great amount of pressure pushing against the seat when the valve is being closed.

Butterfly valves have an arrow stamped on the side, indicating the direction of flow through the valve. They must be installed in the proper flow direction, or the seat will not seal and the valve will leak. Three types of butterfly valves include the wafer, wafer lug, and two-flange valve.

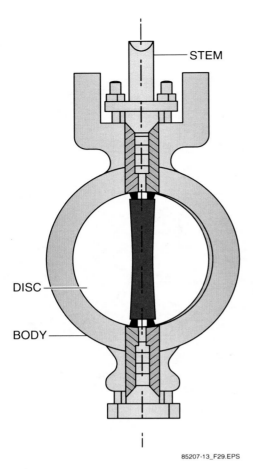

Figure 29 Butterfly valve.

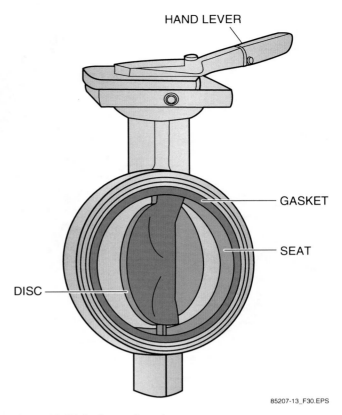

Figure 30 Wafer butterfly valve.

3.3.1 Wafer Valves

The wafer-type butterfly valve (*Figure 30*) is designed for quick installation between two flanges. No gasket is needed between the valve and the flanges because the valve seat is lapped over the edges of the body to make contact with the valve faces. Bolt holes are provided in larger wafer valves to help line up the valve with the flanges. During installation of the valves, the valve must be in the open position prior to torquing to prevent damage to the valve seat.

3.3.2 Wafer Lug Valves

Wafer lug valves (*Figure 31*) are the same as wafer valves except that they have bolt lugs completely around the valve body. Like the wafer valve, no gasket is needed between the valve and the flanges because the valve seat is lapped over the edges of the body to make contact with the valve faces. The lugs are normally drilled to match American National Standards Institute (ANSI) 150-pound steel drilling templates. The lugs on some wafer lug valves are drilled and tapped so that when the valve is closed, downstream piping can be dismantled for cleaning or maintenance

while the upstream piping is left intact. When tapped wafer lug valves are used for pipe end applications, only one pipe flange is necessary.

3.3.3 Two-Flange Valves

The body of a two-flange butterfly valve is made with a flange on each end (*Figure 32*). The valve seat is not lapped over the flange ends of the valve, so gaskets are required between the flanged body and the mating flanges. The two-flange valve is made with either flat-faced flanges or raised-face flanges, and the mating flanges must match the valve flanges.

3.4.0 Diaphragm Valves

A diaphragm valve is one in which flow through the valve is controlled by a flexible disc that is connected to a compressor by a stud molded into the disc. The valve stem moves the compressor up and down, regulating the flow. Diaphragm valves have no seats because the body of the valve acts as the seat for the flexible disc. The operating mechanism of the valve does not come in contact with the material within the pipeline.

These valves can be used fully opened, fully closed, or for throttling service. They are good for handling slurries, highly corrosive materials, or materials that must be protected from contamina-

---

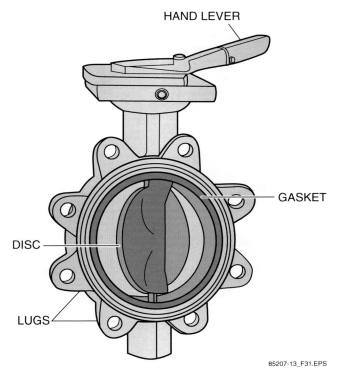

Figure 31 Wafer lug butterfly valve.

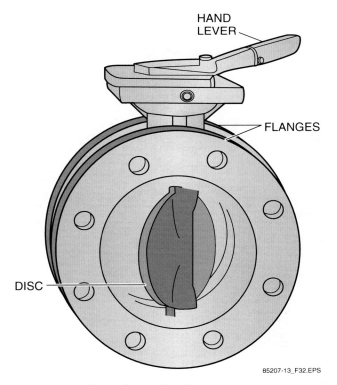

Figure 32 Two-flange butterfly valve.

tion. Many types of fluids that would clog other types of valves will flow through a diaphragm valve. *Figure 33* shows a weir-type diaphragm valve. In addition to the types shown, there is a straight-through type. A special type of diaphragm valve is the pinch valve, in which the flexible lining of the opening is pinched closed by moving bars at the top and bottom of the valve. The linings are replaceable, and prevent exposure of the metal parts of the valve to the substance passing through.

3.5.0 Needle Valves

Needle valves, as shown in *Figure 34*, are used to make relatively fine adjustments in the amount of fluid allowed to pass through an opening. The needle valve has a long, tapering, needle-like point on the end of the valve stem. This needle acts as a disc. The longer part of the needle is smaller than the orifice in the valve seat. It therefore passes through it before the needle seats. This arrangement permits a very gradual increase or decrease in the size of the opening and thus allows a more precise control of flow than could be obtained with an ordinary globe valve. Needle valves are often used as component parts of other, more complicated valves. For example, they are used in some types of reducing valves. Most constant-pressure pump governors have needle valves to minimize the effects of fluctuations in pump discharge pressure. Needle valves are also used in some components of automatic combustion control systems where very precise flow regulation is necessary.

4.0.0 CONTROL VALVES

Control valves (*Figure 35*) are variations of angle or globe valves that are controlled by pneumatic, electronic, or hydraulic actuators. They operate in a partially open position and are most commonly used for pressure limiting or flow control. Control valves are precision-built for increased accuracy in flow control. The control valve is usually smaller than the pipeline to avoid throttling and constant wear of the valve seat. The actuators for control valves are explained in more detail later in the module.

Control valves are often used with pressure-sensing elements that measure the pressure drop at a given point in the pipeline. The flow rate is directly related to the pressure drop. The sensing element sends a pressure signal to a controller that compares the pressure drop with the pressure drop for the desired flow rate. If the actual flow is different, the controller adjusts the valve through the valve actuator to increase or decrease the flow. *Figure 36* shows a schematic for a control valve in a pipeline.

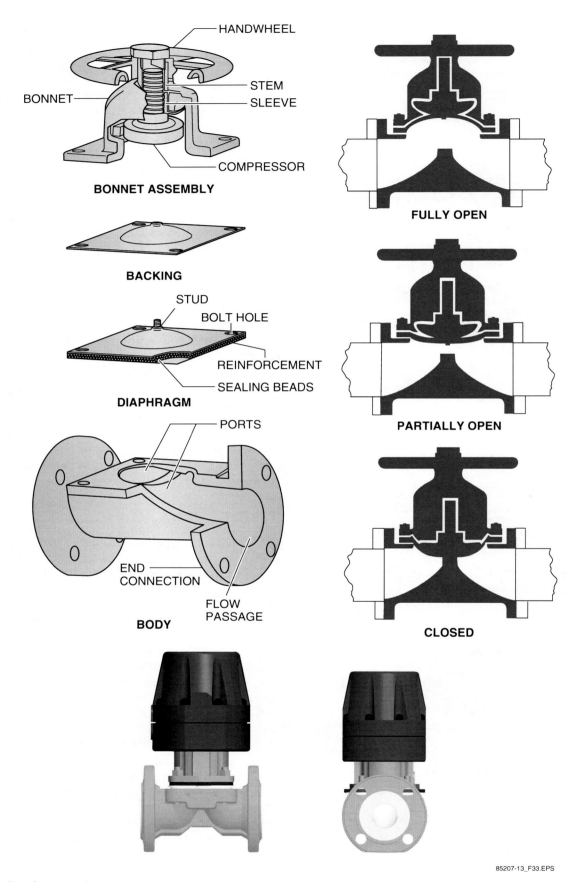

Figure 33 Diaphragm valve.

85207-13_F33.EPS

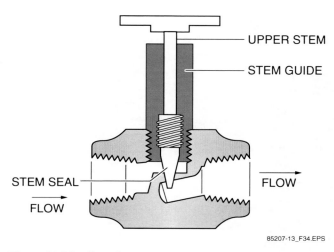

Figure 34 Needle valve.

5.0.0 VALVES THAT RELIEVE PRESSURE

Valves used to relieve pressure in a pipeline, tank, or vessel are known as safety valves and pressure-relief valves (*Figure 37*). These are installed in pipelines and vessels to prevent excess pressure from rupturing the line and causing an accident. Both types are adjustable and operate automatically.

| CAUTION | Proper direction of flow is critical. The wrong orientation can result in damage to the valve. |

5.1.0 Safety Valves

Safety valves are normally used in steam, air, or other gas service pipelines. They operate in the closed position until the pressure in the line rises above the preset pressure limit of the valve. At this point, the valve opens fully to relieve the pressure and remains open until the pressure drops, at which point the valve snaps shut. Because of this fully open and tight-closing feature, safety valves are commonly referred to as pop-off valves. *Figure 38* shows a cutaway view of a safety valve.

Safety valves are most commonly installed in the vertical position. For air or gas service, these valves can be installed upside down to allow moisture to collect, which seals the surfaces. Safety valves should be mounted directly to the tank or vessel being protected using connecting piping and block valves. Before the safety valve can close, its discharge pressure must drop several pounds below the opening pressure. A safety valve has an adjustment to control the closing of the valve, depending on the amount of pressure drop after the valve has opened. This adjustment is calibrated by the manufacturer and should not be adjusted in the field.

5.2.0 Pressure-Relief Valves

Relief valves are normally used to relieve pressure in liquid services. They are operated by pressure acting directly on the bottom of a disc that is held on its seat by a spring. The amount of spring pressure on the disc is determined at the factory when the valve is manufactured. When liquid pressure becomes great enough to overcome the force of the spring on the disc, the disc rises, allowing the liquid to escape. After enough liquid has escaped to allow the pressure to drop, the system pressure becomes too low to overcome the force of the spring on the disc. The spring then pushes the disc back onto its seat, stopping the flow of liquid through the valve. The force of the spring is called the compression of the valve. *Figure 39* shows a pressure-relief valve.

Spring-loaded relief valves are most commonly installed with the stem in the vertical position. The piping connected to these valves must be as large or larger than the valve inlet and outlet openings. A reducer can be used to reduce down to the inlet of a relief valve, but the line size cannot be increased to the inlet of the relief valve. The piping must also be well supported so that the line strains will not cause the valve to leak at the seat.

6.0.0 VALVES THAT REGULATE DIRECTION OF FLOW

Check valves regulate flow by preventing the backflow of liquids or gases in a pipeline. Check valves are intended to permit the flow in only one direction, which is indicated by an arrow stamped on the side of the valve. The force of flow and the action of gravity cause the valve to open and close. Check valves are manufactured in several different designs, including the following:

- Swing
- Lift
- Ball
- Butterfly
- Foot

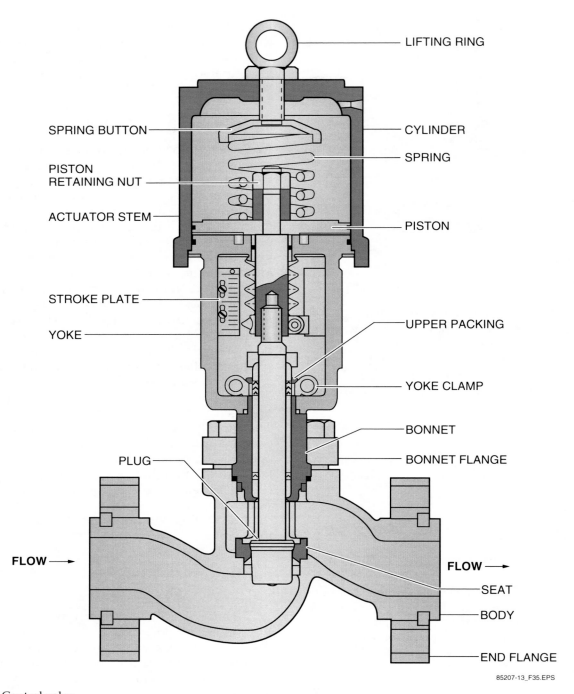

LIFTING RING

SPRING BUTTON

CYLINDER

SPRING

PISTON RETAINING NUT

ACTUATOR STEM

PISTON

STROKE PLATE

YOKE

UPPER PACKING

YOKE CLAMP

BONNET

PLUG

BONNET FLANGE

FLOW →

FLOW →

SEAT

BODY

END FLANGE

85207-13_F35.EPS

Figure 35 Control valve.

> **NOTE**
>
> In a maritime application, swing check valves must be installed in the fore and aft direction. They should never be installed in the port to starboard direction.

6.1.0 Swing Check Valves

The most commonly used type of check valve is the swing check valve (*Figure 40*). This valve has a disc that swings, or pivots, to open and close the valve. Since check valves provide straight-line flow through the valve body, they offer less resistance to flow than other types of check valves. Swing check valves can be mounted horizontally or vertically with the flow upward. If this valve is used in a vertical flow, a counterweight is needed to make sure it does not stick open.

The disc, which is hinged at the top, seats against a machined seat in the tilted wall bridge opening. The disc swings freely from the fully closed position to the fully open position, which provides unobstructed flow through the valve. The flow through the valve causes the disc to stay open, and

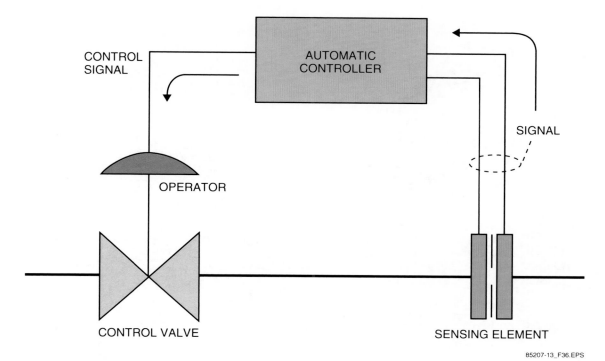

Figure 36 Control valve schematic.

85207-13_F36.EPS

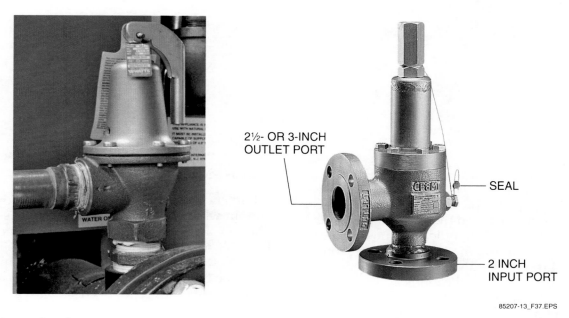

2½- OR 3-INCH
OUTLET PORT

SEAL

2 INCH
INPUT PORT

85207-13_F37.EPS

Figure 37 Examples of pressure-relieving valves.

the amount that the disc is open depends on the volume of liquid moving through the valve. Gravity or reversal of flow causes the disc to close, preventing backflow through the valve.

Swing check valves can also be equipped with an outside lever and counterweight to assist the valve in closing. These valves are often referred to as weighted check valves. A special case of this valve is the flap valve, used in some slurry and waste pipe-to-tank applications to prevent backflow. The flap valve does not have an enclosing

housing, only a flap and a seat. It is essentially the self-closing end of the pipe line.

6.2.0 Lift Check Valves

The lift check valve (*Figure 41*) is designed like a globe valve, with an indirect line of flow through the valve. The lift check valve also works automatically by line pressure and prevents the reversal of flow through the valve. A lift check valve contains a disc that is held against the disc seat

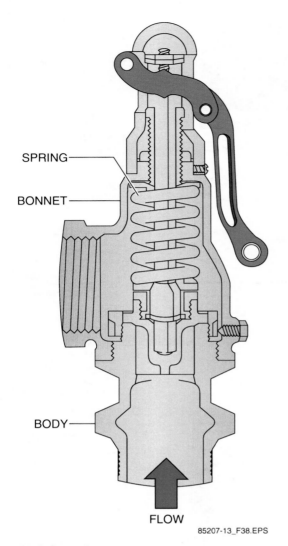

SPRING

BONNET

BODY

FLOW

85207-13_F38.EPS

Figure 38 Safety valve.

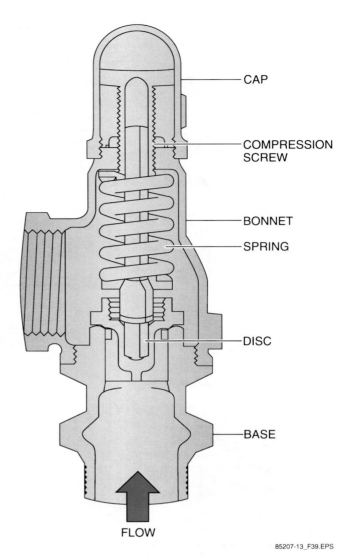

CAP

COMPRESSION SCREW

BONNET

SPRING

DISC

BASE

FLOW

85207-13_F39.EPS

Figure 39 Pressure-relief valve.

by gravity or by the pressure of the fluid flowing in the opposite direction through the valve. When fluid flows through the valve in the correct direction, the force of the fluid lifts the disc away from the seat to allow the fluid to pass through the valve. Lift check valves can be designed for horizontal or vertical applications and must only be used for the application they are designed for. They cannot be installed with the bonnet down in a horizontal installation.

6.3.0 Ball Check Valves

Ball check valves (*Figure 42*) operate in the same manner as lift check valves, except they use a ball instead of a disc to allow the fluid to pass through the valve. Ball check valves that are not spring-loaded should only be installed in the vertical position. Gravity and the reversal of flow cause the ball to rest against the valve seat, restricting flow through the valve.

Ball check valves are recommended for use in lines in which the fluid pressure changes rapidly since the action of the ball is practically noiseless. The ball rotates during operation and therefore tends to wear at a uniform rate over its entire surface. The ball valve also stops flow reversal more rapidly than other types of check valves.

6.4.0 Butterfly Check Valves

A butterfly check valve has two vanes that resemble the wings of a butterfly. These vanes fold back from a central hinge to open and allow flow but close to prevent backflow. *Figure 43* shows a butterfly check valve.

> **NOTE**
>
> If the butterfly check valve is used in the horizontal position, it must have a spring to assist in closing.

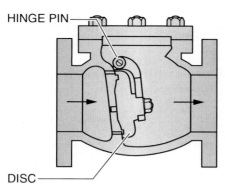

85207-13_F40.EPS

Figure 40 Swing check valves.

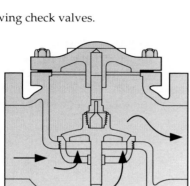

HORIZONTAL LIFT CHECK

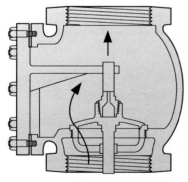

VERTICAL LIFT CHECK

85207-13_F41.EPS

Figure 41 Lift check valve.

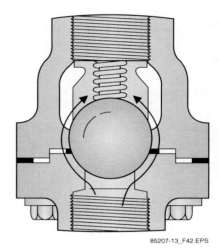

85207-13_F42.EPS

Figure 42 Ball check valve.

6.5.0 Foot Valves

Foot valves (*Figure 44*) are used at the bottom of the suction line of a pump to maintain the prime of the pump. This type of valve operates similarly to the lift check valve. On the inlet side of the foot valve is a strainer that keeps foreign material out of the line. The weight of the liquid in the pipeline between the pump and the foot valve keeps the suction pipeline full when the pump is shut down. The weight pushes the seats closed in the valve, preventing the liquid from flowing

85207-13_F43.EPS

Figure 43 Butterfly check valve.

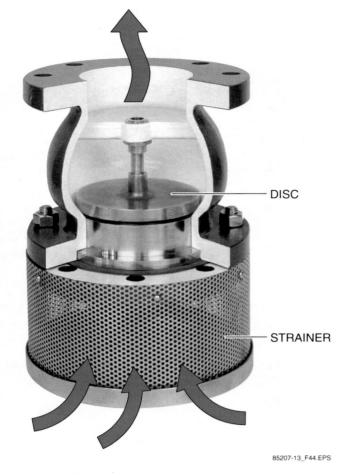

DISC

STRAINER

85207-13_F44.EPS

Figure 44 Foot valve.

out through the valve. When the pump starts, the force of the suction causes the liquid coming into the valve to push against the outside of the seats to open the valve.

> **NOTE**
>
> If the foot valve is used in the horizontal position, it must have a spring to assist in closing.

7.0.0 VALVE ACTUATORS

The primary purpose of a valve actuator, also known as an operator, is to provide automatic control of a valve or to reduce the effort required to manually operate a valve. A standard handwheel or hand lever attached to a valve stem is one type of actuator commonly used on smaller valves that do not require great effort to operate. Larger valves must be equipped with some other type of actuator. Hard-to-reach valves are sometimes equipped with a type of extension or remote operator commonly called a **reach rod**.

Many valves are mechanically operated and powered by electric, pneumatic, or hydraulic ac-

tuators or operators. Electric, pneumatic, and hydraulic valve actuators are used when a valve is remotely operated; when the frequency of operation of the valve would require unreasonable human effort; or if rapid opening or closing of the valve is required. There are many types of valve actuators in use today, including the following:

- Gear operators
- Chain operators
- Pneumatic and hydraulic actuators
- Electric or air motor-driven actuators

7.1.0 Gear Operators

Gear operators minimize the effort required to operate large valves that work at unusually high pressures. Gear operators are also used to connect to valves located in inaccessible areas. Three basic types of gear operators are spur-gear operators, bevel-gear operators, and worm-gear operators.

7.1.1 Spur-Gear Operators

Spur gears are used to connect parallel shafts. They have straight teeth cut parallel to the shaft axis and use gears to transfer motion and power from one shaft to another parallel shaft. Spur-gear operators can be as simple as one gear transmitting power to another gear attached to the valve stem, or they can consist of many gears, depending on how much power needs to be transmitted. *Figure 45* shows a spur-gear operator.

Although there may be valves with open gear trains in older installations, this design is no longer permitted by OSHA. Gear-driven valves in current use have enclosed gear boxes like the one shown in *Figure 46*.

7.1.2 Bevel-Gear Operators

Bevel gears are used to transmit power between two shafts that intersect at a 90-degree angle. The bevel gear resembles a cone, in that the teeth on each gear are angled at 45 degrees to mesh with the gears on the mating gear. Bevel-gear operators transmit power from a handwheel and a shaft that is perpendicular to the valve stem (*Figure 47*).

7.1.3 Worm-Gear Operators

Worm gears are used in butterfly valves to transmit power at right angles (*Figure 48*). This means that the shafts of the connecting gears are at 90-degree angles to each other. The worm is a cylindrical-shaped gear similar to a screw, and is the driver. The worm gear is the larger, circular gear in the assembly. Worm-gear operators transmit

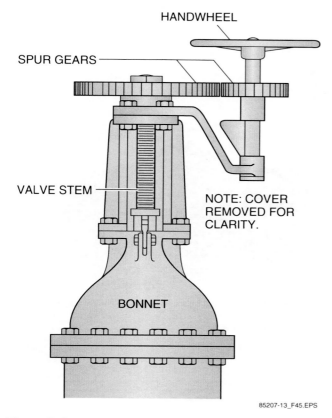

Figure 45 Spur-gear operator.

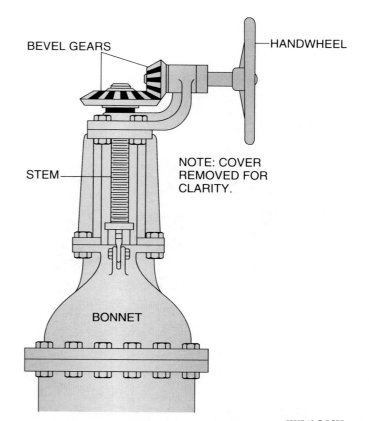

Figure 47 Bevel-gear operator.

Figure 46 Enclosed gear box valve.

power from the handwheel, which is attached to the worm, to the worm gear, which is attached to the valve stem. Worm gears are set up like bevel gears, with a side-mounted handwheel. A high-speed ratio is obtainable with a worm-gear operator because the operator allows rapid opening and closing of large valves with minimum effort. It takes only one-quarter turn of the handwheel to open or close the valve.

7.2.0 Chain Operators

A chain operator (*Figure 49*) is installed in some situations when the valve is mounted too high to reach the handwheel. The stem of the valve is mounted with a chain wheel, and the chain is brought to within 3 feet of the working floor level.

Chain operators are used only when they are absolutely necessary. Universal-type chain wheels that attach to the regular valve handwheel have been blamed for many industrial accidents. In corrosive atmospheres where an infrequently operated valve is located, the attaching bolts of this type of chain wheel have been known to fail. Chain wheels attach to the stem and replace the regular valve handwheel.

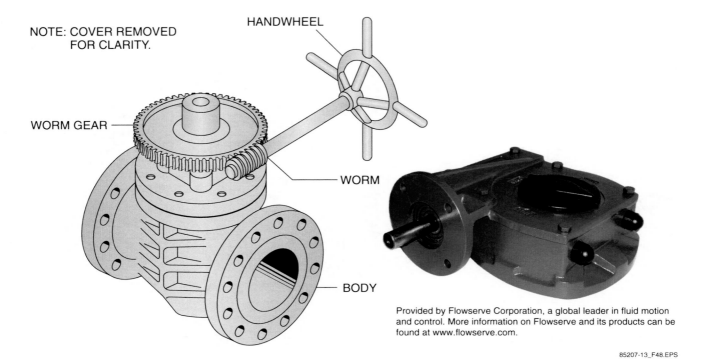

NOTE: COVER REMOVED FOR CLARITY.

HANDWHEEL

WORM GEAR

WORM

BODY

Provided by Flowserve Corporation, a global leader in fluid motion and control. More information on Flowserve and its products can be found at www.flowserve.com.

85207-13_F48.EPS

Figure 48 Worm-gear operator.

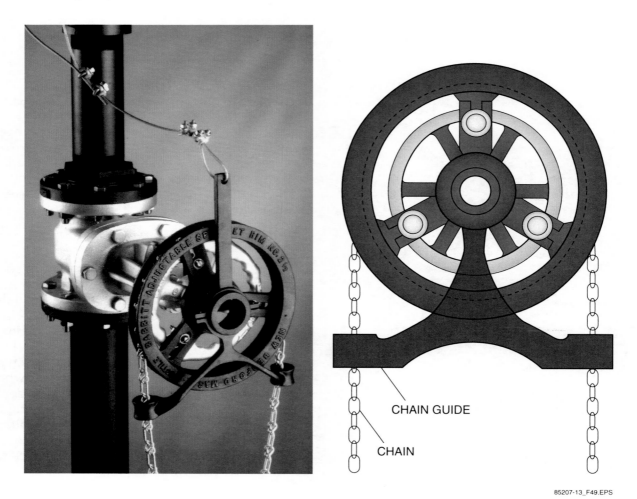

CHAIN GUIDE

CHAIN

85207-13_F49.EPS

Figure 49 Chain operator.

OSHA now requires that the wheels for chain operators be secured with a safety cable (lanyard). This is done to prevent them from falling if their fasteners have been weakened by rust or corrosion.

7.3.0 Pneumatic and Hydraulic Actuators

Pneumatic and hydraulic actuators operate in basically the same manner, except the pneumatic actuator operates off air pressure and the hydraulic actuator operates off fluid pressure. A cylinder assembly that contains a piston is attached to the valve stem (*Figure 50*). Air or fluid pressure above and below the piston moves the piston and the valve stem up and down, opening and closing the valve. Many pneumatic and hydraulic actuators also have spring-loaded pistons in either the naturally open or naturally closed position, depending on which position is considered the fail-safe position. These springs allow the valve to fail open or fail closed. If the control medium is lost, the spring will force the valve to the selected fail-safe position, which is determined as part of the system design process. This safety feature is required so that the valve will return to its safe position in case of pressure failure. *Figure 51* shows a spring-loaded pneumatic valve actuator.

WARNING!

The springs in pneumatic valve actuators are under strong tension and can cause serious injury if let loose. Technicians must not disassemble a valve unless they have been trained for that work.

7.4.0 Electric or Air Motor-Driven Actuators

Electric or air motor-driven valve actuators contain a motor that is linked through reduction gears to the valve stem. These are known as motor-actuated valves, or MOVs. The electric motor is equipped with electrical limit switches that shut off the motor when the motor has turned the valve stem as far as it can go in either direction. This prevents unnecessary wear and tear on the motor. Electric valve actuators are usually equipped with a handwheel for controlling the valve manually if the power fails. Actuators should only be removed, turned, and adjusted by qualified technicians.

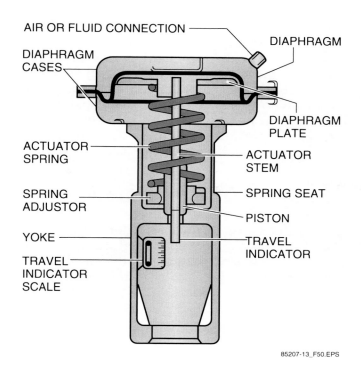

Figure 50 Actuator components.

AIR- OR FLUID-TO-LOWER ACTUATOR

AIR- OR FLUID-TO-RAISE ACTUATOR

85207-13_F51.EPS

Figure 51 Spring-loaded pneumatic valve actuators.

7.5.0 Reach Rods

In maritime applications, a valve might be located in a hard-to-reach or hazardous location. One way to safely operate a valve in such a situation is to use an extension device called a reach rod. Many reach rods, such as the one shown in *Figure 52*, are flexible so that they can be routed through tight spaces to operate valves that might ordinarily be inaccessible. For instance, on a ship, valves might be located deep inside the hull. In order to enable the valves to be operated efficiently, reach rods might be used to extend control through a floor or bulkhead.

8.0.0 STORING AND HANDLING VALVES

Once a valve has been received at the job site, it must be properly stored until it is installed in the pipeline. If the valve has been abused in storage, it may fail after installation. Valves must be handled carefully to avoid damaging them.

8.1.0 Safety Considerations

When working with and around valves, maritime pipefitters must be alert and work cautiously to ensure their own safety and the safety of their co-workers. Follow these guidelines when handling or working around valves:

- Be aware of all pinch points.
- Do not stand under a load.
- Watch for overhead power lines and other equipment.
- Do not overload temporary work platforms.

Figure 52 Flexible reach rod.

- Ensure that final supports and hangers are in place before installing a large valve.
- Never operate any valve in a live system without proper authorization.
- Always use a spud wrench or drift pin when aligning bolt holes in a flanged valve. Never align bolt holes with your fingers.
- Never stand in front of a safety valve relief discharge.

8.2.0 Storing Valves

Most valve manufacturers wrap the valves in protective wrapping for shipment. This wrapping should remain on the valve until installation. Follow these guidelines when storing and handling valves:

- Clearly label all valves with identification tags or stenciled placards.
- Store all valves on appropriate hardwood dunnage to keep the valves off the ground. Never store valves on the ground.
- Store all valves on the basis of their compatibility with other valves. For example, do not store stainless steel valves with carbon steel valves because the rust from carbon steel can cause the stainless steel to corrode.
- Store valves in an area where corrosive fumes, freezing weather, and excessive water can be kept to a minimum.
- Store the valves in an area where no objects can fall and damage them.
- Do not remove any tags from valves.
- Ensure that all open ends have end protectors.
- Store valve handles with their mating valves.
- When storing valves outside, always lay them on their side so that water cannot get trapped inside the valve. If water inside the valve freezes, it can shatter the body material.

8.3.0 Rigging Valves

Special precautions must also be taken when rigging valves. Follow these guidelines when rigging valves:

- Clean and protect all threads and weld ends before lifting.
- Select rigging equipment based on the weight of the valve.
- Do not rig a valve by the stem, handle, actuator, or through the body opening. Rig only to the valve body.
- Rig stainless steel valves using nylon straps only.
- Install a tag line to control the lift.

- Rig for the proper position of the valve in the final installation.
- Rig to allow for installation of bolts and nuts.
- Remove all shipping materials before installation.

9.0.0 VALVE PLACEMENT

Proper placement of valves is very important to the efficient operation of the piping system. Valves can be installed in piping systems with welded, threaded, or flanged joints. The procedure for installing a valve is the same for installing a fitting. However, there are added factors to consider when installing valves.

The location of a valve in a piping system is very important. Most piping drawings indicate exactly where to install the valve. The direction in which the stem and handwheel are to be located when working with small-bore piping is often the pipefitter's responsibility. When working with large-bore piping, the stem direction and orientation are normally shown on the drawings. Because of regular maintenance procedures, the valve must be easily accessible when possible. The height of the valve must be accessible without causing hazards. The best installation height for valve handles is between 2 feet and 4 feet 6 inches off the floor or working level. Valves installed below and above this area (up to 6 feet 6 inches) create either a tripping hazard or a face hazard; therefore, the valve handwheel must have some type of guard around it. *Figure 53* shows the order of preference for valve location.

Valves that are installed with the stem in the upright position tend to work best. The stem can be rotated down to the horizontal position, but should not point down. If the valve is installed with the stem in the downward position, the bonnet acts like a trap for sediment, which may cut and damage the stem. Also, if water is trapped in the bonnet in cold weather, it may freeze and crack the body of the valve.

Another factor that must be considered when installing valves is the direction of flow through the valve. Butterfly, safety, pressure-relief, and some other valves either have arrows indicating direction of flow stamped on the side, or the ports are labeled as the inlet or the outlet. When the valve is not marked, determine which side of the disc the pressure flows against. Gate valves

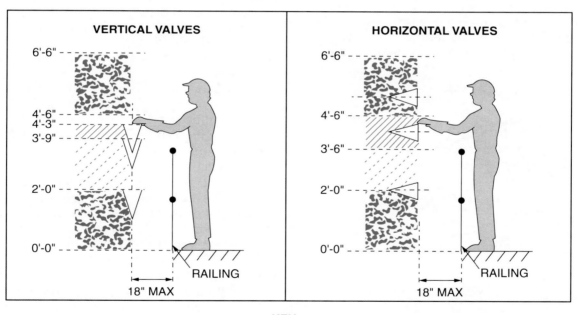

KEY

⬛ PREFERRED ELEVATIONS

⬛ SECOND-CHOICE ELEVATIONS

⬛ LEG OR HEAD HAZARD, UNLESS PROTECTION GIVEN BY RAILING, PIPING, OR EQUIPMENT

85207-13_F53.EPS

Figure 53 Order of preference for valve location.

can have the pressure on either side. Globe valves should be installed so that the pressure is below the disc unless pressure above the disc is required in the job specifications (*Figure 54*).

10.0.0 VALVE SELECTION, TYPES, AND APPLICATIONS

Because of the diverse nature of valves, with valve types overlapping each other in both design and application, the valve selection process must be examined. This section discusses valve selection, valve types, and valve applications.

10.1.0 Valve Selection

With valve selection, there are many factors that must be considered. Cost is often an overriding factor, although experience has shown that sparing expense now may result in additional expense later. When selecting a valve during system design, overall system performance must be considered. Questions that must be asked include the following:

- At what temperature will the system be operating? Are there any internal parts that would be adversely affected by the temperature? Valves designed for high temperature steam systems

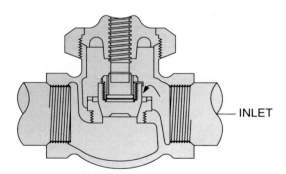

PRESSURE ABOVE DISC

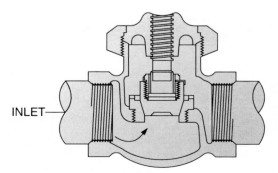

PRESSURE BELOW DISC

85207-13_F54.EPS

Figure 54 Pressure above and below disc.

are not necessarily suited for the extreme low temperatures that may be found in, say, a liquid nitrogen system.

- At what pressure (or vacuum) will this valve be operating? How does the temperature affect the valve's pressure rating? System integrity is a major concern on any system. The valve must be rated at or above the maximum system pressure anticipated. Due to factors such as valve design, packing construction, and end attachments, the valve is often considered a weak point in the system.

- Are there any sizing constraints? It seems obvious that a 2-inch valve would not be installed in a 10-inch pipe, but what may not be obvious is how the yoke size, actuator, or **positioner** figures in the scenario. Valve manufacturers provide dimensional tables to aid in valve selections.

- Will this valve be used for an on-off or a throttling application? Throttle valves are generally globe valves, although in some applications a ball valve or butterfly valve may be used.

- To what type of erosion will the valve be exposed? Will it require hardened seats and discs? Will it be throttled close to its seat and need a special pressure drop valve?

- What kind of pressure drop is allowed? Globe valves exhibit the largest pressure drop or head loss characteristics, whereas ball valves exhibit the least.

- What kind of differential pressure will this valve be operated against? Will this differential pressure be used to seat or unseat the valve? Will the high differential pressure deform the body or disc and bind the valve? Will this also require a bypass valve?

- How will this valve be connected to the system? Will it be welded, screwed, or flanged? Should it be butt welded or socket welded? Will it be union threaded or pipe threaded? Should the flanges be raised, flat, **phonographic**, male/female, or tongue-and-groove?

- In what type of environment will the valve be installed? Is it a dirty environment where an exposed stem would score the **yoke bushing** and cause premature failure? Is it a clean environment where a different stem lubricant should be used?

- What kind of fluid is being handled? Is it hazardous in such a way that packing leakage may be detrimental? Is it corrosive to the packing or to the valve itself?

- What is the life expectancy required? Will it require frequent maintenance? If so, is it easily repairable, or does the cost of labor justify replacement instead of repair?

If an installed valve is to be replaced, a valve identical to the one removed should be installed. If that valve is no longer manufactured, valve selection should be made in the same fashion as for a new application. The valve dimensions are the limiting factor unless piping alterations can be made. Several questions should be asked:

- Are the system parameters the same as when the system was designed, or has the system intent changed?
- Have any problems been noted since system fabrication that could be remedied by installing a different valve design at this time?
- With what type of operator should the new valve be fitted? Is the new valve compatible with the installed operator?

10.2.0 Valve Types and Applications

As noted in the valve selection discussion, there are many factors that determine the application and/or type of valve to be used. Some of the factors include:

- The temperature at which the system will be operating
- The sizing constraints or the pressure at which the system will be operating
- The kind of fluid or material that is in the system
- The type of environment in which the valve will have to operate
- The type of actuator the valve will use

11.0.0 VALVE MARKINGS AND NAMEPLATE INFORMATION

Before the present system of valve and flange coding, manufacturers had their own systems. With the development of components rated at higher temperatures and pressures, in conjunction with more stringent regulations, a standard was needed. The Manufacturers Standardization Society (MSS) first developed SP-25 in 1934. In 1978, SP-25 was revised to incorporate all the changes that had developed since 1934. To preclude errors in cross-referencing, the American National Standards Institute (ANSI) and the **ASTM International** have adopted the MSS marking system.

Two markings that are frequently used on valves are the flow direction arrow, indicating which way the flow is going, and the bridgewall marking, shown in *Figure 55*. The bridgewall marking is usually found on globe valves and is an indication of how the seat walls are angled in

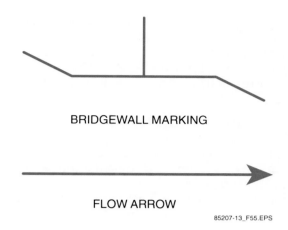

BRIDGEWALL MARKING

FLOW ARROW

85207-13_F55.EPS

Figure 55 Valve markings.

relation to the inlet and outlet ports of the valve. Specifically, it shows whether the wall of the seat on the inlet side angles up or down. The wall of the seat on the outlet side of the globe valve will always be angled opposite to the angle on the inlet side, as indicated in *Figure 56*.

Not all globe valves are designed with angled bridgewalls. However, some applications may specifically require the process to enter either on the top side or the bottom side of the disc in a globe valve.

If the process enters on the top side (bridgewall angled up on the inlet), the force of the process will assist in the closing of the valve. However, if the process enters on the bottom side (bridgewall angled down on the inlet), the force of the process will assist in the opening of the valve.

Markings for flow are not normally used on gate, plug, butterfly, or ball valves. If a gate valve has a flow arrow, it is because the gate valve has a double gate. Double-gate valves are capable of relieving fluid pressure in the event that a high pressure difference exists across the shut gate. Standard practice is for the outlet-side gate to relieve to the inlet side. This type of valve is used for specific applications. Therefore, system plans should be consulted for correct valve orientation.

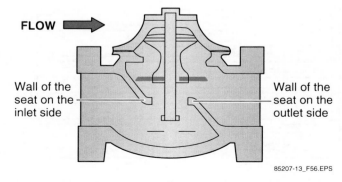

FLOW

Wall of the seat on the inlet side

Wall of the seat on the outlet side

85207-13_F56.EPS

Figure 56 The meaning of bridgewall markings on valves.

There are normally two identification sets: one permanently embossed, welded, or cast into the valve body, and the other a valve identification plate (*Figure 57*). Typically, as a minimum, the following information will be included within the two sets:

- Rating designation markings
- Material designation markings
- Melt identification markings
- Trim identification markings (if applicable)
- Size markings
- Thread identification markings (if applicable)

The omission of markings and component marking requirements must also be discussed.

11.1.0 Rating Designation

The rating designation of a valve gives the pressure and temperature rating as well as the type of service. *Table 2* shows commonly used service designations.

The product rating may be designated by the class numbers alone, as with a steam pressure rating or pressure class designation. The ratings for products that conform to recognized standards, but are not suitable for the full range of pressures or temperatures of those standards, may be marked as appropriate. The numbers and letters representing the pressure rating at the limiting conditions may also be shown.

The rating designation for products that do not conform to recognized national product standards may be shown by numbers and letters representing the pressure ratings at maximum and minimum temperatures. If desired, the rating designation may be shown as the maximum pressure followed by cold working pressure (CWP) and the allowed pressure at the maximum temperature (for example, 2,000 CWP 725/925°F).

Other typical designations are given as the first letter of the system for which they are designated:

- A – Air service
- G – Gas service
- L – Liquid service
- O – Oil service
- W – Water service
- DWV – Drainage, waste, and vent service

11.2.0 Trim Identification

Trim identification marking is required on the identification plate for all flanged-end and butt welding end steel or flanged-end ductile iron body valves with trim material that is different than the body material. Symbols for materials are the same. The identification plate may be marked with the word trim followed by the appropriate material symbol.

Trim identification marking for gate, globe, angle, and cross valves, or valves with similar design characteristics, consists of three symbols. The first indicates the material of the stem. The second indicates the material of the disc or wedge face. The third indicates the material of the seat face. The symbol may be preceded by the words stem, disc, or seat, or it may be used alone. If used alone, the symbols must appear in the order given.

Plug valves, ball valves, butterfly valves, and other quarter-turn valves require no trim identification marking unless the plug, disc, closure member, or stem is of different material than the body. In such cases, trim identification symbols on the nameplate first indicate the material of the stem and then the material of the plug, ball, disc, or closure member.

Those valves with seating or sealing material different than the body material must have a third symbol to indicate the material of the seat. In these cases, symbol identification must be preceded by the words stem, disc (or plug, ball, or gate, as appropriate) and the word seat. If used alone, the symbols must appear in the order given.

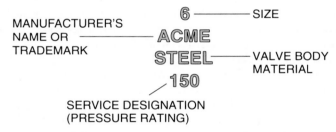

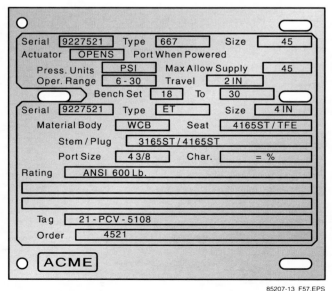

Figure 57 Valve nameplate.

Table 2 Valve Service Designations

Correspond to Steam Working Pressure (SWP)	Correspond to Cold Working Pressure (CWP)
Steam pressure (SP)	Water, oil pressure (WO)
Working steam pressure (WSP)	Oil, water, gas pressure (OWG)
Steam (S)	Water, oil, gas pressure (WOG)
	Gas, liquid pressure (GLP)
	Working water pressure (WWP)
	Water pressure (WP)

85207-13_T02.EPS

11.3.0 Size Designation

Size markings are in accordance with the product-referenced marking requirements. For size designation for products with a single nominal pipe size of the connecting ends, the word *nominal* indicates the numerical identification associated with the pipe size. It does not necessarily correspond to the inside diameter of the valve, pipe, or fitting.

Products with internal elements that are the equivalent of one pipe size, or are different than the end size, may have dual markings unless otherwise specified in a product standard. Unless these exceptions exist, the first number indicates the connecting end pipe size. The second indicates the minimum bore diameter, or the pipe size corresponding to the closure size (for example, 6×4, $4 \times 2\frac{1}{2}$, 30×24).

At the manufacturer's option, triple marking size designation may be used for valves. If triple size designation is used, the first number must indicate the connecting end size at the other end. For example, $24 \times 20 \times 30$ on a valve designates a size 24 connection, a size 20 nominal center section, and a size 30 connection.

Fittings with multiple outlets may be designated at the manufacturer's option in a run × run × outlet size method. For example, $30 \times 30 \times 24$ on a fitting designates a product with size 30 end connections and a nominal size 24 connection between.

11.4.0 Thread Markings

Fittings, flanges, and valve bodies with threaded connecting ends other than American National Standard Pipe Thread or American National Standard Hose Thread will be marked to indicate the type. The style or marking may be the manufacturer's symbol provided confusion with standard symbols is avoided. Fittings with left-hand threads must be marked with the letters LH on the outside wall of the appropriate opening.

Marking of products with ends threaded for American Petroleum Institute (API) casing, tubing, or drill pipe must include the nominal size, the letters API, and the thread type symbol as listed in *Table 3*.

Marking of products using other pipe threads must include the following:

- Nominal pipe, tubing, drill pipe, or casing size
- Outside diameter of pipe, tubing, drill pipe, or casing
- Name of thread
- Number of threads per inch

11.5.0 Valve Schematic Symbols

The last and most important aspect of valve identification is the ability to identify different types of valves from drawings and schematics. In general, the symbols that denote various control valves, actuators, and positioners are standard symbols as shown in *Figure 58*. However, in certain cases these symbols vary, depending on site-specific prints. The legend of a typical system print or schematic will show the symbols that represent all components on the drawing.

Table 3 Examples of Threaded-Type Symbols

Name/Description	Symbol
Casing – Short round thread	CSG
Casing – Long round thread	LCSG
Casing – Buttress thread	BCSG
Casing – Extreme-line	XCSG
Line pipe	LP
Tubing – Non-upset	TBG
Tubing – External-upset CSG	UP TBG

85207-13_T03.EPS

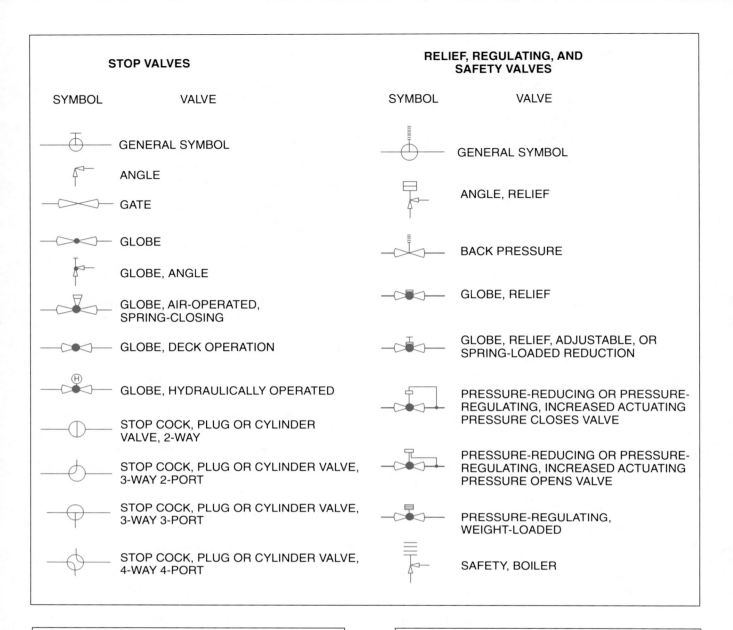

STOP VALVES

SYMBOL   VALVE

GENERAL SYMBOL

ANGLE

GATE

GLOBE

GLOBE, ANGLE

GLOBE, AIR-OPERATED,
SPRING-CLOSING

GLOBE, DECK OPERATION

GLOBE, HYDRAULICALLY OPERATED

STOP COCK, PLUG OR CYLINDER
VALVE, 2-WAY

STOP COCK, PLUG OR CYLINDER VALVE,
3-WAY 2-PORT

STOP COCK, PLUG OR CYLINDER VALVE,
3-WAY 3-PORT

STOP COCK, PLUG OR CYLINDER VALVE,
4-WAY 4-PORT

**RELIEF, REGULATING, AND
SAFETY VALVES**

SYMBOL   VALVE

GENERAL SYMBOL

ANGLE, RELIEF

BACK PRESSURE

GLOBE, RELIEF

GLOBE, RELIEF, ADJUSTABLE, OR
SPRING-LOADED REDUCTION

PRESSURE-REDUCING OR PRESSURE-
REGULATING, INCREASED ACTUATING
PRESSURE CLOSES VALVE

PRESSURE-REDUCING OR PRESSURE-
REGULATING, INCREASED ACTUATING
PRESSURE OPENS VALVE

PRESSURE-REGULATING,
WEIGHT-LOADED

SAFETY, BOILER

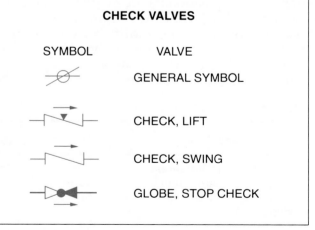

CHECK VALVES

SYMBOL   VALVE

GENERAL SYMBOL

CHECK, LIFT

CHECK, SWING

GLOBE, STOP CHECK

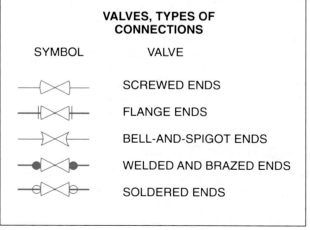

**VALVES, TYPES OF
CONNECTIONS**

SYMBOL   VALVE

SCREWED ENDS

FLANGE ENDS

BELL-AND-SPIGOT ENDS

WELDED AND BRAZED ENDS

SOLDERED ENDS

85207-13_F58.EPS

Figure 58 Typical piping system schematic symbols.

12.0.0 Flanges

Flanges are used to connect valves to manage other piping system components. A flanged joint consists of two flanges, a gasket or other device to seal the joint, and the necessary bolts or stud bolts to hold the joint together. Flanges are considered to be mechanical parts that can be assembled and disassembled repeatedly and with relative ease.

Like most other piping components, flanges are made from several different types of materials. The flanges are available in several different basic primary ratings. The primary rating is a pressure rating based on a pressure/temperature relationship. *Table 4* shows the basic primary ratings. The material used in flanges must match the pipe material.

12.1.0 Types of Flanges

Flanges are used in different types of maritime piping systems and different service applications. Some of the most common flanges include the following:

- Weld neck
- Slip-on
- Slip-on reducing
- Blind
- Socket weld
- Threaded
- Lap-joint
- Silver-brazed

12.1.1 Weld Neck Flanges

Weld neck flanges (*Figure 59*) are distinguished from other types of flanges by a long, tapered hub that is welded to the pipe end. This hub reinforces the flange and allows material to flow smoothly between the pipe and the flange. This transition between flange thickness and pipe wall thickness is very helpful under severe conditions, such as repeated bending caused by thermal expansion, and strengthens the endurance of the flange assemblies to equal that of a typical butt weld pipe joint. The

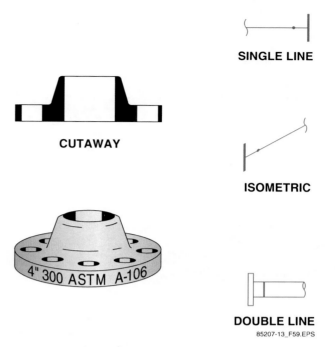

Figure 59 Weld neck flange.

weld neck flange is preferred over other flanges for very **severe service** conditions. Weld neck flanges must be marked with the weight, as in standard, extra heavy, or extra-extra heavy. The markings will be stamped on the neck of the flange. The flanges of different weights have different bore sizes, that is, internal diameters, as do the pipe of the different weights, and the bores would not match if the pipe and flanges were not matched.

12.1.2 Slip-On Flanges

Slip-on flanges are preferred by many companies because of the low initial cost, reduced accuracy required in cutting the pipe to length, and greater ease in alignment procedures. Slip-on flanges slip over the end of the pipe and are fillet-welded to the pipe both at the neck and at the face of the flange. Normal practice is to extend the flange face beyond the end of the pipe about ⅜ inch or the wall thickness of the pipe, whichever is greater.

Slip-on flanges (*Figure 60*) are used where the space is limited and a weld neck cannot be used and are also used with many low-pressure applications of Class 150 to Class 300.

12.1.3 Slip-On Reducing Flanges

The slip-on reducing flange (*Figure 61*) is used for reducing the line size when a weld neck flange and reducer combination will not fit. A slip-on reducing flange should be used only when the flow is from the smaller-sized pipe to the larger-sized pipe. If the flow is in the opposite direction, the smaller

Table 4 Basic Primary Ratings

Forged Steel		Cast Iron
Class 150	Class 900	Class 25
Class 300	Class 1,500	Class 125
Class 400	Class 2,500	Class 250
Class 600		Class 800

85207-13_T04.EPS

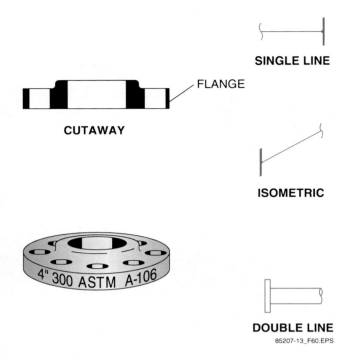

SINGLE LINE

FLANGE

CUTAWAY

ISOMETRIC

DOUBLE LINE

85207-13_F60.EPS

Figure 60 Slip-on flange.

opening will cause turbulence or an excessive **pressure differential** at that point. If the difference in sizes of the standard and reducing flanges is too great, the pressure differential can produce **cavitation**. What actually happens is that the liquid enters the smaller pipe and speeds up, and loses pressure. When the pressure drops below the vapor pressure of the liquid (the pressure at which it becomes a gas instead of a liquid) bubbles form. When the bubbles collapse, the result is a physical impact wave to the vessel or parts of the vessel. This makes the waves keep striking the inside of the pipe, or the fitting to which it is attached, and damages the surface. If space limitations make a slip-on reducing flange necessary, the approval of the equipment manufacturer must be obtained. Manufacturers of equipment such as pumps, heaters, and heat exchangers must control the pressure drops in their systems, and the equipment nozzle sizes directly affect the pressure drop through the system. This is why it is not uncommon for equipment nozzle sizes to differ from pipe sizes. Slip-on reducing flanges are available in all sizes and ratings.

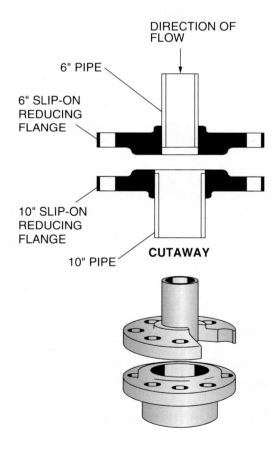

DIRECTION OF FLOW

6" PIPE

6" SLIP-ON REDUCING FLANGE

10" SLIP-ON REDUCING FLANGE

10" PIPE

CUTAWAY

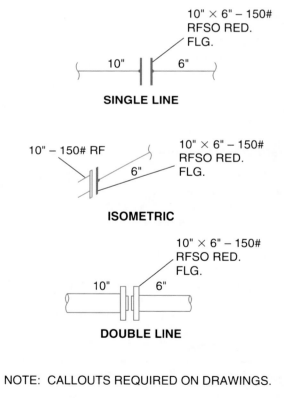

10" × 6" – 150# RFSO RED. FLG.

10"   6"

SINGLE LINE

10" – 150# RF

6"

10" × 6" – 150# RFSO RED. FLG.

ISOMETRIC

10" × 6" – 150# RFSO RED. FLG.

10"   6"

DOUBLE LINE

NOTE: CALLOUTS REQUIRED ON DRAWINGS.

85207-13_F61.EPS

Figure 61 Slip-on reducing flange.

12.1.4 Blind Flanges

Blind flanges (*Figure 62*) are used to close off the ends of pipes, valves, and **pressure vessel** openings. From the standpoint of internal pressure and bolt loading, blind flanges are the most highly stressed of all flange types since the maximum stresses in a blind flange are bending stresses at the center. These flanges are available in the seven primary ratings, and their dimensions can be found in manufacturer's catalogs.

12.1.5 Socket Weld Flanges

Socket weld flanges (*Figure 63*) are attached to pipe ends by inserting the pipe into a socket on the flange and then fillet-welding the flange to the pipe at the flange hub. These flanges are used primarily with small, high-pressure piping. Socket weld flanges are available in Class 150, Class 300, Class 600, and Class 1,500 pressure ratings.

12.1.6 Threaded Flanges

Threaded flanges are available in all sizes and pressure ratings up to and including 12 inches and Class 2,500. However, the use of high-pressure threaded flanges should be limited to special applications. Their major advantage is that they can be assembled without welding, but the threaded joint will not withstand high temperatures or bending stresses. *Figure 64* shows a threaded flange.

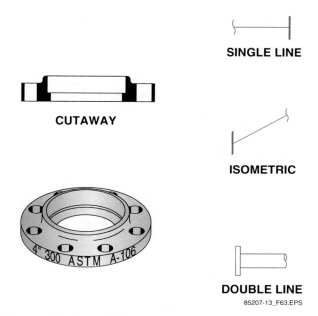

Figure 63 Socket weld flange.

12.1.7 Lap-Joint Flanges

Lap-joint flanges, also known as Van Stone flanges, are used with lap-joint stub ends. Lap-joint stub ends are straight pieces of pipe with a lap on the end, which the flange rests against. These can be used at all pressures, and the flange is normally not welded to the stub end, making alignment easier. When using lap-joint flanges in vertical applications, have a welder weld steel nubs to the sides of the stub end beneath the flange. This prevents the flange from sliding down the pipe when the bolts are being removed.

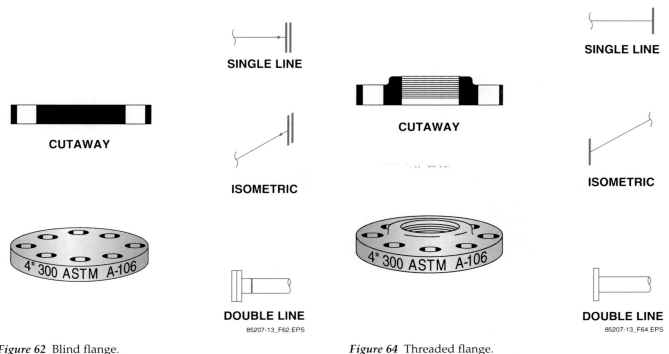

Figure 62 Blind flange.

Figure 64 Threaded flange.

The main use of lap-joint flanges in carbon steel piping systems is in services that require frequent disassembly for inspection and cleaning. When used in systems subject to high corrosion, these flanges can usually be salvaged because they do not actually come in contact with the substance flowing through the pipe.

These flanges are also used in many cases for economic reasons. When running stainless steel piping systems, you can use carbon steel lap-joint flanges with stainless steel stub ends to save money. The use of lap-joint flanges at points where severe bending stresses occur should be avoided. *Figure 65* shows a lap-joint flange and a stub end.

12.1.8 Silver-Brazed Flanges

Silver-brazed flanges, commonly known as silbraze flanges, typically have a pipe-size end socket with precision groove designed to accommodate a silver brazing ring insert. The ring insert, which is ordered separately, allows them to be joined without the use of additional silver solder.

12.2.0 Flange Facings

Facing refers to the sealing surface machined onto flanges. Flange facings are governed by standards to ensure that all flanges of each type and rating have the same facing. Typically, two different face styles at one joint are never used at the same time, but this may be necessary when connecting pipe

to manufactured equipment. The industry standard for flange facing finish is the phonographic serrated finish. The phonographic finish means that the facing of the flange has a continuous narrow, shallow spiral groove in the face. The other common facing finish is a concentric serrated finish. This is a series of shallow circular grooves sharing a common center. The normal range is between 30 and 55 grooves per lateral inch of the diameter of the face. Both finishes allow soft gasket material to fill the grooves and keep liquids from leaking. Some special industrial applications may use a smooth finish, sometimes called a cold water finish, to be used without a gasket. The various types of flange facings include the following:

- Raised-face
- Flat-face
- Ring joint type
- Tongue and groove
- Male and female

Figure 66A shows flange facing finishes and *Figure 66B* shows flange facings.

12.2.1 Raised-Face Flanges

The raised-face flange has a wide, raised rim around the center of the flange and is the most common facing used with steel flanges. The raised face on Class 150 and Class 300 flanges is $\frac{1}{16}$ inch high, and the face on flanges of all other pressure ratings is $\frac{1}{4}$ inch high. This flange uses a wide, flat ring gasket that is squeezed tightly be-

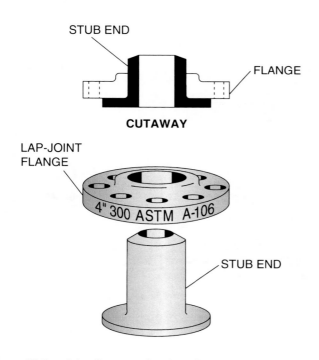

Figure 65 Lap-joint flange and stub ends.

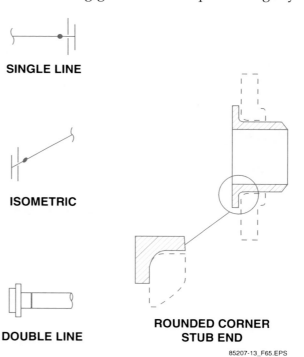

85207-13_F65.EPS

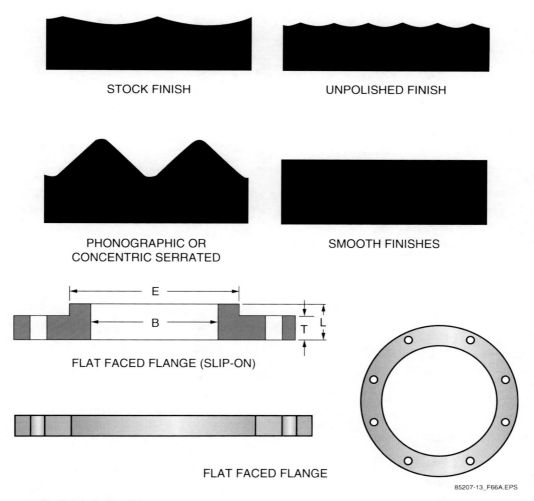

STOCK FINISH

UNPOLISHED FINISH

PHONOGRAPHIC OR CONCENTRIC SERRATED

SMOOTH FINISHES

FLAT FACED FLANGE (SLIP-ON)

FLAT FACED FLANGE

85207-13_F66A.EPS

Figure 66 Flange facing finishes. (1 of 2)

tween the flanges. The facing is machine-finished with tiny, spiral grooves that bite into and help hold the gasket.

12.2.2 Flat-Face Flanges

The flat-face flange is flat across the entire face. This flange uses a gasket that has bolt holes cut in it and that covers the entire flange face. Flat-face flanges are mainly used to join Class 150 and Class 300 forged steel flanges with Class 125 and Class 250 cast iron flanges. The brittle nature of cast iron flanges requires that flat-faced flanges be used so that full face contact is achieved between the flanges.

If a flat-face flange must be connected to a raised-face flange, it must be confirmed that the flat-face flange is not cast iron because the pressure of the bolts can crack or break the cast iron flange. If the flat-face flange is cast iron, the raised-face flange must be changed to a flat-face flange only after obtaining engineering approval.

12.2.3 Ring Joint Type Flanges

The ring joint type (RJT) flange (*Figure 67*) has a different type of facing because the contact surface of the seal is below the actual flange facing. This type of flange uses an oval, rectangular, or octagonal steel ring as a seal that fits into a ring groove that is machined into the faces of the flanges. Only one steel ring is required for each pair of mating flanges.

The RJT flange facing is the most expensive standard facing, but for some oil and refinery applications, it is usually regarded as the most efficient because internal pressure acts on the ring to increase the sealing force. One disadvantage of using this type of flange is that in order to remove a valve or a section of pipe from the system, the flanges have to be spread far enough apart to remove the ring. In tight hookups, this can cause problems.

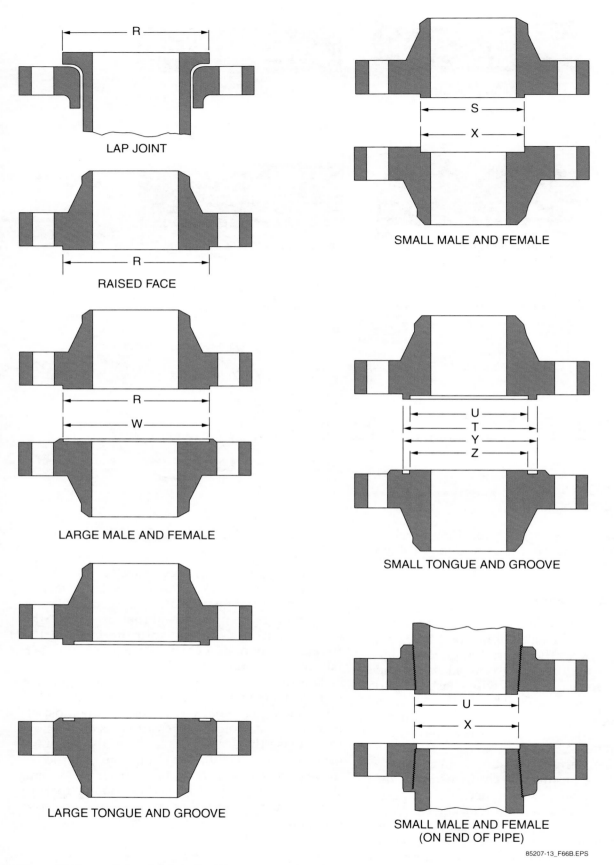

LAP JOINT

RAISED FACE

LARGE MALE AND FEMALE

LARGE TONGUE AND GROOVE

SMALL MALE AND FEMALE

SMALL TONGUE AND GROOVE

SMALL MALE AND FEMALE
(ON END OF PIPE)

85207-13_F66B.EPS

Figure 66 Flange facing finishes. (2 of 2)

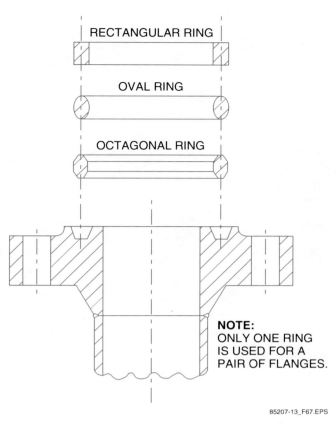

RECTANGULAR RING

OVAL RING

OCTAGONAL RING

NOTE:
ONLY ONE RING
IS USED FOR A
PAIR OF FLANGES.

85207-13_F67.EPS

Figure 67 Metal ring joint gaskets.

12.2.4 Tongue and Groove Flanges

Tongue and groove flange faces are classified as either large tongue and groove or small tongue and groove. This type of flange was originally developed for hydraulic service and offers high gasket pressure in a small area between the tongue and the groove. One advantage that this type of flange offers is that it is more resistant to flange stresses and bending. Disadvantages include excessive gasket pressure with changes in temperature, mushrooming of the tongue under excessive bolt torquing, and the tight tolerances involved in machining and assembly.

12.2.5 Male and Female Flanges

Male and female flange faces fit together much like the tongue and groove flange facings. A rim on the male face fits into a machined groove in the female flange. A flat ring gasket is used between the rim and the groove in the female flange. The same disadvantages that apply to tongue and groove flanges also apply to male and female flange facings, and male and female flanges are also classified as either large or small.

13.0.0 FLANGE GASKETS

Gaskets are available in a variety of types and materials (*Figure 68*). The type of gasket used must be matched to the process characteristics and operating conditions to which it will be exposed. For example, different gasket materials are capable of withstanding different temperature and pressure ranges, and certain gasket materials are compatible with different process fluids. In piping systems, gaskets are placed between two flanges to make the joint leak proof. Gaskets are normally sized by the thickness of the gasket material in fractions of an inch.

To ensure that the composition of various types of gaskets, gasket material, and some packing meets the required specifications, ASTM International devised a method of identifying different types of gasket material.

In a maritime environment, pipe fitters might be exposed to commercial and US Navy piping systems. It should be noted, as shown in *Figure 68*, that commercial and military piping flanges and gaskets can differ. For instance, similar flanges might have different numbers and sizes of bolt holes.

Based on manufacturer's specs, adhesive materials such as Loctite® gasket replacement are used in certain applications, especially in machine assemblies, as a replacement for cut gaskets. The material is applied from a tube or pump to the clean surface of the flange, and replaces the gasket. Such applications are medium strength, and the instructions of the manufacturer must be followed closely. The drawbacks include having to use only recommended cleaning solutions beforehand, and that the material takes from 24 hours to a week to cure.

13.1.0 Gasket Materials

Gaskets are made of many different types of materials to meet the demands of the particular process system in which they are installed. Generally, one of the following four conditions will exist in a process system. These conditions affect the types of gaskets that can be used in the system.

- *High temperature/low pressure* – Generally includes temperatures from 500°F to 1,200°F (260°C to 649°C) and pressures up to 600 psi.
- *High temperature/high pressure* – Generally includes temperatures from 500°F to 1,200°F (260°C to 649°C) and pressures from 600 to 2,500 psi.

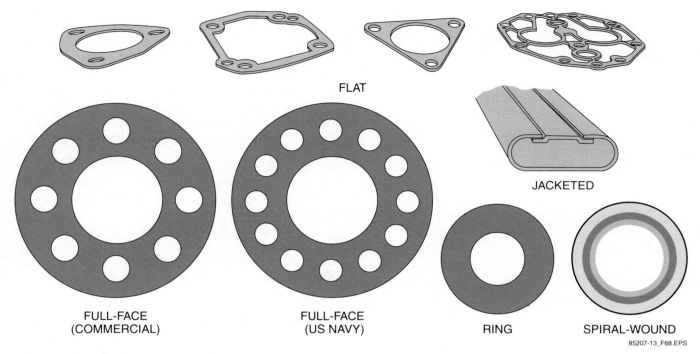

FLAT

JACKETED

FULL-FACE
(COMMERCIAL)

FULL-FACE
(US NAVY)

RING

SPIRAL-WOUND

85207-13_F68.EPS

Figure 68 Gaskets.

- *Low temperature/low pressure* – Generally includes temperatures up to 500°F (260°C) and pressures up to 600 psi.
- *Low temperature/high pressure* – Generally includes temperatures up to 500°F (260°C) and pressures from 600 to 2,500 psi.

Usually, the types of gaskets and gasket materials to be used on a project are listed in a special section of the piping specifications. Gasket materials are rated by ANSI specifications for applicable pressures.

Since most process systems are subject to a combination of temperature and pressure, it is extremely important to match the proper gasket material to the operating conditions of the process system. The type of gasket material used is based on several factors. Generally, the gasket is chosen to withstand the temperature, pressure, and chemical properties of the process medium. Different gasket materials have different pressure and temperature ratings. These ratings are a reflection of the ability of a gasket material to withstand forces in a process system. Various types of gaskets include the following:

- PTFE (Teflon®)
- Fiberglass
- Acrylic fiber
- Metal
- Rubber
- Cork
- Vinyl
- Ceramic

- Ethylene propylene diene monomer (EPDM)
- Neoprene
- Nitrile
- Silicone
- Viton®
- Gylon

13.1.1 PTFE Gaskets

PTFE flange gaskets, also known as Teflon®, are nonporous, nongalvanic gaskets. They are usually white and are available in several pressure/temperature limits. PTFE is available in different forms and products. Standard amorphous PTFE, the original form of PTFE, is available in sheet form and can be cut into the various sizes necessary to fit the applications. Amorphous PTFE is frequently associated with a phenomenon referred to as gasket creep and cold flow (creep) which results in a loss of bolt load. When under continuous load, PTFE tends to continuously flow due to a lack of strength. This often creates a necessity for plant personnel to retorque bolts in order to compensate for this material flow.

Fillers may introduce other problems, including different characteristics from the non-stick and chemically stable PTFE. The flow was slowed somewhat, but other advantages were lost.

Finally a process was discovered to expand PTFE in a manner so as to dramatically reduce and nearly eliminate PTFE gasket creep and cold flow. Products are made from virgin PTFE resin, have no fillers, have the same chemical compatibility as the original PTFE, operate in the same temperature range (−450°F to +500°F) (232°C to 260°C), and have no colorizers or other additives. Expanded PTFE, due to its very high tensile strength properties, can be used in pressure environments from full vacuum to 3,000 psi.

The company that makes GORE-TEX® outerwear also makes cord gasket material (*Figure 69*), made from expanded PTFE, that is used to make gaskets. It is sold in rolls, to be applied to the joint by hand. The biggest advantage of PTFE is that it is chemically inert. This means that it does not react chemically with the process medium.

13.1.2 Fiberglass Gaskets

Fiberglass gaskets are treated so that they will remain soft and pliable. They sometimes include wire inserts for service for boiler handhold and manhole covers, as well as for tank heads and other high-pressure applications. Fiberglass gaskets are usually white in color. They can typically be used in applications up to 380°F (190°C) and 180 psi.

13.1.3 Acrylic Fiber Gaskets

Acrylic fiber flange gaskets are used with oil, steam, weak acids, and alkalines. They are usually off-white in color. Acrylic fiber gaskets can typically be used in applications up to 700°F (371°C) and 1,200 psi. Many gaskets combine fibers of various materials filled with nitrile, EPDM, or neoprene (*Figure 70*).

13.1.4 Metal Gaskets

Metal gaskets are used for machine flanges and where metal-to-metal fits are required. They are good for applications in which higher tempera-

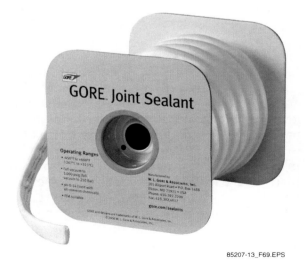

85207-13_F69.EPS

Figure 69 Cord gasket material.

85207-13_F70.EPS

Figure 70 Acrylic fiber gasket.

ture fluctuations and pressures are normal. When using metal gaskets, the material must be suitable for use within the process system. The following are some of the types of metal gaskets available:

- *Solid-metal gaskets* – These gaskets are made from solid metal. They are used in processes suitable for the metal or alloy used to make the gasket. Common materials are copper, Monel, steel, and iron. Copper gaskets are not used in high-pressure high-temperature systems. Monel, steel, and iron gaskets can be used in a variety of pressure and temperature applications.

- *Ring joint metal gaskets* – The four types of standard ring joint gaskets (*Figure 71*), defined by the ring section, are: R octagonal; R oval; RX (eccentric octagonal); and BX (rectangular). The octagonal is considered superior. The rings are made of the softest carbon steel or iron available. For lower temperatures, the rings are made of plastic to resist corrosion or to insulate the joint from electric currents. The ring joint gasket is effective at very high pressures.

- *Serrated metal gaskets* – Serrated metal gaskets are solid metal gaskets that have concentric ribs machined into their surfaces. With the contact area reduced to a few concentric lines, the required bolt load is reduced considerably. This design forms an efficient joint. Serrated gaskets are used with smooth-finished flange faces.

- *Corrugated metal gaskets* – Corrugated metal gaskets are generally used on low-pressure systems where the flanges are smooth and bolt pressure is low. The ridges of the corrugations

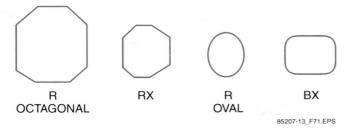

R
OCTAGONAL   RX   R
                              OVAL   BX

85207-13_F71.EPS

Figure 71 Standard ring joint gasket types.

tend to concentrate the gasket loading along the concentric ridges.

- *Laminated metal gaskets* – These gaskets are made of metal with a soft filler. The laminate can be parallel to the flange face or spiral-wound. Laminated gaskets can be used in high-pressure, high-temperature applications.
- *Corrugated metal gaskets with asbestos inserted* – These gaskets are used for a variety of pressure and temperature applications in steam, water, gas, air, oil, oil vapor, and refrigerant systems.
- *Flat metal jacket gaskets with asbestos gasket material* – These gaskets are used in the same applications as corrugated metal gaskets with asbestos inserted.
- *Corrugated metal jacket gaskets with heat-resistant, synthetic filler* – These gaskets are used for the same applications as corrugated metal with asbestos-inserted gaskets. Synthetic replacements for asbestos are often referred to as nonasbestos.
- *Spiral-wound metal gaskets with heat-resistant, synthetic filler* – Spiral-wound metal gaskets are used for a variety of pressure and temperature applications in the same type systems as corrugated metal jacket gaskets with heat-resistant, synthetic filler. All spiral-wound metal gaskets are color-coded to identify the type of metal from which they are made. Spiral-wound gaskets normally have an inner ring, an outer ring, and a section of preformed windings in the middle that serve as the sealing element. The inner and outer rings are usually spot-welded, and serve as blowout preventers for the sealing element. The preformed shapes of the sealing element are not welded, and are usually slightly thicker through the gasket.

Table 5 shows the industry-standard color code for spiral-wound gaskets.

13.1.5 Natural Rubber Gaskets

Natural rubber is seldom used in installations today. It has low solvent and oil resistance, but is excellent when used with water. Natural rubber has excellent resilience and may be used at tem-

Table 5 Industry-Standard Color Code for Spiral-Wound Gaskets

Metal	Color
Type 304 SS	Yellow
Type 316 SS	Green
Type 347 SS	Blue
Type 321 SS	Turquoise
Monel	Orange
Nickel	Red
Titanium	Purple
Alloy 20	Black
Carbon steel	Silver
Hastelloy B	Brown
Hastelloy C	Beige
Inconel	Gold
Filler Material	**Color of Stripe**
Polytetrafluoroethylene (PTFE)	White
Ceramic	Light green
Flexible graphite	Gray
Mica/graphite	Salmon

85207-13_T05.EPS

peratures up to 175°F (79°C). Rubber gaskets are available in a variety of pressure and temperature ratings. Rubber gaskets generally come in thicknesses from ¹⁄₁₆ to 1 inch. Rubber gaskets are used in low-pressure/low-temperature water, gas, air, and refrigerant systems. The following types of rubber gaskets are used:

- *Standard black or red rubber gaskets* – These gaskets are used for saturated steam up to 100 psi. They have an approximate temperature range of –20 to 170°F (-29 to 77°C).
- *Reinforced rubber gaskets* – These gaskets are strengthened by polyester fabric plies. They are commonly used for saturated steam and low-pressure steam. Reinforced rubber gaskets have an approximate temperature range of –40 to 200°F (-40 to 93°C). They are typically black. In addition to fabric reinforcement, nylon monofilament is used to reinforce rubber gaskets.
- *High-test, reinforced rubber gaskets* – These gaskets are used for pressures up to 500 psi. They have an approximate temperature range of –40°F to 200°F (-40 to 93°C), and they are black.

13.1.6 Cork Gaskets

Cork gaskets are compressible, flexible, light-weight, resilient, and nonabsorbent. They resist most oils, petroleum products, and chemicals. Cork is rarely used in industry, except in special cases, as a vibration-absorbing material because of its cellular construction. It can be used for most liquids, even when boiling, except for strong alkalis. Cork gaskets are generally available in thicknesses from ¹⁄₁₆ to ½ inch.

13.1.7 Vinyl Gaskets

Vinyl gaskets are specially fabricated for use as oil-resistant gaskets. They also have a good resistance to water, chemicals, oxidizing agents, ozone, and abrasion. Vinyl gaskets are black in color and have an approximate temperature range of 20°F to 160°F (–7 to 71°C).

13.1.8 Ceramic Gaskets

Ceramic gaskets are used for high-temperature air applications, such as boiler systems, where other types of materials would fail due to the high temperature. Ceramic gaskets can typically be used for temperatures exceeding 1,500°F (815°C).

13.1.9 Asbestos Gaskets

Some gaskets contain encapsulated asbestos. These gaskets are used in high-temperature applications. Inform a supervisor if the metallic exterior is broken and the white asbestos fibers exposed.

In most locations, legal requirements exist for any action that may cause any possible exposure to asbestos. The area where the asbestos is to be removed must be completely isolated from the outside air. Even a single fiber of asbestos in a worker's lungs has the potential to cause a form of lung cancer. Asbestos can only be removed by properly trained and certified workers. In addition to an airtight seal, asbestos workers are required to wear protective clothing and an air supply, all of which is decontaminated before leaving the sealed area. If the gasket needs to be replaced, another type of gasket must be used.

WARNING!

If you do find an asbestos gasket, do not attempt to deal with it. Inform your supervisor immediately.

13.1.10 EPDM

EPDM is very good in an oxygen environment. It has good ozone resistance. EPDM has low resistance to solvent and oil. It is a good material to be used with water and chemicals and has a maximum temperature rating of 350°F (177°C). It also has excellent flame resistance.

13.1.11 Neoprene

Neoprene, which has good resilience, is only used in noncritical conditions. Neoprene has fair oil and solvent resistance, but has poor chemical resistance and is not recommended for water service. It has a maximum temperature rating of 250°F (121°C).

13.1.12 Nitrile

Nitrile is used with medium-pressure oil and solvent services. Nitrile cannot be used with acetone or methyl ethyl ketone (MEK). Nitrile has a maximum temperature rating of 250°F (121°C). Nitrile has poor ozone resistance, but is an excellent material to use with water or alcohol. Nitrile also has good chemical resistance but poor flame resistance.

13.1.13 Silicone

Silicone is a rubber-like material that is widely used. There are more grades of silicone material than any other rubber-type material. It has a maximum temperature rating of 550°F (288°C). Silicone is not recommended for use with oil or solvents. It has fair to good chemical resistance, excellent ozone resistance, and fair to good flame resistance. It has poor to excellent resilience depending on the grade used. It is not recommended for use on water systems.

13.1.14 Viton®

Viton® is a fluoroelastomer material. It has excellent chemical resistance and is the most commonly used chemical-resistant material. It has a maximum temperature rating of 400°F (204°C). Viton®, a hard material, has excellent oil and solvent resistance and also has excellent ozone resistance. Viton® is a relatively inexpensive gasket material.

13.1.15 Gylon® or Amerilon®

Under proprietary names like Gylon® and Amerilon®, there are also some new materials that combine Kevlar® and PTFE, to produce a very tough and resistant low temperature gasket. It is chemically inert, has a high resilience (45 percent recovery), and a maximum temperature rating of 500°F (260°C). Gylon® 3510 has one of the best sealability factors of any gasket material in the industry. It is one of the most commonly used gasket materials. It is also a good choice for mating dissimilar metals.

13.1.16 Graphite-Impregnated Gaskets

These gaskets are made of fiberglass and impregnated with graphite particles. They are used to provide a tight seal in some high-temperature applications. However, they are reactive to oxidation.

13.1.17 Soft Metal Gaskets

Soft metal and aluminum gaskets are designed to compress into the concentric rings machined into the surface of carbon steel and stainless steel flanges. Metal gaskets are used for oil and gas pipelines, where their non-reactive materials and high wear resistance keep them working over long periods without replacement.

13.1.18 Gasket Color Codes

Manufacturers use color codes to help identify gaskets made of different materials. These color codes will vary from one manufacturer to another. Some manufacturers use an additional color-coding scheme on the outside diameter of some gaskets so the material can be identified without breaking the seal. In such cases, the gasket body may be the same color as other types of gaskets, but the OD color will be different. The important thing is to check the color-coding scheme used by the given manufacturer before replacing a gasket.

> **NOTE**
>
> The use of a commercial liquid gasket remover can make removing old gasket material easier. It reduces the amount of scraping and sanding needed, thus preventing possible damage to flange surfaces.

13.2.0 Types of Flange Gaskets

There are different types of gaskets that must be matched to different types of flanges. For example, flat ring gaskets are used with raised-face flanges; full-face gaskets are used with flat-face flanges; and metal ring joint gaskets are used with RJT flanges.

13.2.1 Flat Ring Gaskets

Flat ring gaskets are used on flanges with raised faces. They have an outside diameter slightly larger than the outside diameter of the raised face. The material used may be metallic, nonmetallic, or a combination of both. The raised-face flange dimensions and class rating of the flange determine the size of flat ring gasket needed. *Figure 72* shows a flat ring gasket.

13.2.2 Full-Face Gaskets

Full-face gaskets are used with flat-face flanges and extend all the way to the outer edge of the flange. These gaskets have bolt holes that match the holes of the flanges with which they are used. *Figure 73* shows a full-face gasket.

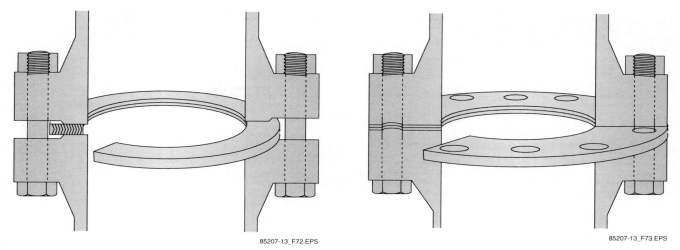

Figure 72 Flat ring gasket.

85207-13_F72.EPS

Figure 73 Full-face gasket.

85207-13_F73.EPS

SUMMARY

Piping systems use valves to start and stop flow, regulate flow, relieve pressure, and prevent reversal of flow. The type of valve used depends on the type of piping system, the nature of the fluid in the system, the temperature and pressure of the fluid in the system, and the desired operation of the system.

Actuators and positioners are energy transmission devices that move the valve stem. They may use any of several different energy sources to perform their functions. These devices provide a means through which valves may be operated remotely or against extremely high differential pressures.

Flanges provide a way to join together sections of piping, especially in systems that might need to be disassembled occasionally. A flanged joint has two flanges, a gasket or other device to seal the joint, and the bolts or studs needed to hold the joint together. As with valves and actuators, flanges and flange gaskets are available in many different sizes and materials.

An integral part of a maritime pipefitter's job is to recognize and understand the many different types of valves, actuators, flanges, and gaskets used in piping systems. Part of that job involves understanding the factors that must be considered during valve, flange, and gasket selection. The information presented in this module should prove invaluable to a maritime pipefitter who is involved in selecting and installing any of these components in a fluid system.

1. Which of the following is correct with regard to gate valves?

   a. They are used to throttle fluid flow.
   b. They open in registered increments.
   c. They are used to stop or start fluid flow.
   d. They are not designed to provide a tight seal.

2. Which of the following valves is a rotary-action valve?

   a. Gate
   b. Needle
   c. Globe
   d. Ball

3. A globe valve that closes against the direction of fluid flow is said to be _____.

   a. flow-to-open
   b. flow-to-close
   c. fail-to-open
   d. fail-to-close

4. What type of valve is most likely to be used together with a pressure-sensing element and a controller to regulate the flow rate of liquid through a pipe?

   a. Pressure-relief valve
   b. Control valve
   c. Knife gate valve
   d. Three-way valve

5. From these choices, select the statement that is true of a typical safety valve.

   a. It is used in liquid service piping and vessel applications.
   b. If system pressure rises above a setpoint, the valve will throttle open and closed until it is locked shut.
   c. The normal operating position of the valve is partially open.
   d. If system pressure rises above the preset limit, the valve will open fully.

6. A swing check valve requires a counter-weight if used in a _____.

   a. horizontal application
   b. vertical application
   c. steam application
   d. liquid application

7. Which of these devices is most likely to be located on the inlet side of a foot valve?

   a. A counterweight
   b. A gate
   c. A strainer
   d. A diaphragm

8. The primary purpose of a valve actuator is to _____.

   a. provide automatic control or reduce the effort required to manually operate a valve
   b. allow valve maintenance without having to shut down the system
   c. enable a typical on-off valve to function in a throttling mode
   d. maintain system pressure by minimizing turbulence and cavitation

9. Which of these valve actuators has straight teeth cut parallel to the shaft axis and uses gears to transfer motion and power from one shaft to another parallel shaft?

   a. Worm-gear operator
   b. Spur-gear operator
   c. Wafer-gear operator
   d. Bevel-gear operator

10. A valve actuator that uses a cylindrical-shaped gear that is similar to a screw to drive a circular gear that is attached to the valve stem is called a _____.

    a. bevel-gear actuator
    b. worm-gear actuator
    c. ring-gear actuator
    d. spur-gear actuator

11. On what type of valve actuator does OSHA require a safety cable (lanyard) to prevent the operator wheel from falling off due to corrosion?

    a. Pneumatic actuator
    b. Hydraulic actuator
    c. Chain operator
    d. Electrical handwheel

12. What feature do electric valve actuators usually have for controlling the valve manually if the power fails?

   a. Pneumatic piston
   b. Hydraulic cylinder
   c. Chain sprocket
   d. Handwheel

13. What are control valves in hard-to-reach or hazardous locations sometimes equipped with to enable personnel to operate the valve?

   a. Reach rod
   b. Bonnet
   c. Slip-on flange
   d. Butterfly

14. From these choices, select the statement that is true about storing valves prior to installation.

   a. Always store valves flat on the ground to prevent damage from falling.
   b. Keep valve ends open to prevent moisture buildup and rusting.
   c. Store all valves in the same place, regardless of their type or composition.
   d. Label all valves with some type of identification tag or placard.

15. When a valve is being prepared for installation, any rigging that is used should be attached _____.

   a. directly to the stem to prevent tangling
   b. only to the valve body
   c. through the valve body opening
   d. to the valve's handle or actuator

16. The best installation height for valve handles is _____.

   a. as close to the floor or working level as possible
   b. any height above 2 feet off the floor or working level
   c. between 2 feet and 4 feet 6 inches off the floor or working level
   d. approximately 5 feet 6 inches off the floor or working level

17. If a gate valve has a flow arrow indicated on it, it is because the gate valve has a _____.

   a. single gate
   b. double gate
   c. bridgewall marking
   d. vent port

18. Trim identification marking for gate, globe, angle, and cross valves consists of three symbols—the first indicates the material of the stem, the second indicates the material of the disc or wedge face, and the third indicates the _____.

   a. material of the seat face
   b. pressure range of the valve
   c. piping material compatibility
   d. temperature rating of the valve

19. Fittings that are marked LH on the outside wall indicate _____.

   a. low heat
   b. liquid hydrogen only
   c. low hardening material
   d. left-hand threads

20. A type of piping flange that is recognizable by its long, tapered hub that is welded to the pipe end is a _____.

   a. slip on flange
   b. weld neck flange
   c. blind flange
   d. lap-joint flange

21. To avoid problems like turbulence and cavitation, a slip-on reducing flange should only be used when flow through a system is from a _____.

   a. smaller-sized pipe to a larger-sized pipe
   b. non-pressurized vessel to an outlet
   c. pipe section to a pressure-relief valve
   d. larger-sized pipe to a smaller-sized pipe

22. From the standpoint of internal pressure and bolt loading, the most highly stressed of all flange types are _____.

   a. threaded flanges
   b. socket weld flanges
   c. blind flanges
   d. flat-face flanges

23. Which of these choices is considered to be the industry standard for flange facing finish?

   a. Polished smooth surface
   b. Phonographic or concentric serrated
   c. Octagonal ring joint
   d. Convex tongue and groove

24. Gaskets impregnated with graphite are used primarily in _____.

    a. high-pressure applications
    b. high-temperature applications
    c. low-pressure applications
    d. low-temperature applications

25. Soft metal gaskets are most likely to be used in pipelines carrying _____.

    a. water
    b. caustic chemicals
    c. steam
    d. oil or gas

Trade Terms Introduced in This Module

Actuator: The part of a regulating valve that converts electrical or fluid energy to mechanical energy to position the valve.

ASTM International: Founded in 1898, a scientific and technical organization, formerly known as the American Society for Testing and Materials, was formed for the development of standards on the characteristics and performance of materials, products, systems, and services.

Angle valve: A type of globe valve in which the piping connections are at right angles.

Ball valve: A type of plug valve with a spherical disc.

Body: The main part of the valve. It contains the disc, seat, and valve ports. The body of the valve is directly connected to the piping by threaded, welded, or flanged ends.

Bonnet: The part of a valve containing the valve stem and packing.

Butterfly valve: A quarter-turn valve with a plate-like disc that stops flow when the outside area of the disc seals against the inside of the valve body.

Cavitation: The result of pressure loss in liquid, producing bubbles (cavities) in vapor in liquid.

Check valve: A valve that allows flow in one direction only.

Control valve: A globe valve automatically controlled to regulate flow through the valve.

Corrosive: Causing the gradual destruction of a substance by chemical action.

Deformation: A change in the shape of a material or component due to an applied force or temperature.

Disc: Part of a valve used to control the flow of system fluid.

Elastomeric: Elastic or rubberlike. Flexible, pliable.

Galling: An uneven wear pattern between trim and seat that causes friction between the moving parts.

Gasket: A device that is used to make a pressure-tight connection and that is usually in the form of a sheet or a ring.

Gate valve: A valve with a straight-through flow design that exhibits very little resistance to flow. It is normally used for open/shut applications.

Globe valve: A valve in which flow is always parallel to the stem as it goes past the seat.

Head loss: The loss of pressure due to friction and flow disturbances within a system.

Kinetic energy: Energy of motion.

Packing: Material used to make a dynamic seal, preventing system fluid leakage around a valve stem.

Phonographic: When referring to the facing of a pipe flange, serrated grooves cut into the facing, resembling those on a phonograph record.

Plug: The moving part of a valve trim (plug and seat) that either opens or restricts the flow through a valve in accordance with its position relative to the valve seat, which is the stationary part of a valve trim.

Plug valve: A quarter-turn valve with a ported disc.

Positioner: A field-based device that takes a signal from a control system and ensures that the control device is at the setting required by the control system.

Pressure differential: The difference in pressure between two points in a flow system. It is usually caused by frictional resistance to flow in the system.

Pressure vessel: A metal container that can withstand high pressures.

Reach rod: An extension device, usually flexible, that is attached to a control valve located in a hard-to-reach or hazardous location to enable remote manual control by an operator.

Relief valve: A valve that automatically opens when a preset amount of pressure is exerted on the valve disc.

Seat: The part of a valve against which the disc presses to stop flow through the valve.

Severe service: A high-pressure, high-temperature piping system.

Thermal transients: Short-lived temperature spikes.

Throttling: The regulation of flow through a valve.

Torque: A twisting force used to apply a clamping force to a mechanical joint.

Trim: Functional parts of a pump or valve, such as seats, stem, and seals, that are inside the flow area.

Turbulence: The motion of fluids or gases in which velocities and pressures change irregularly.

Valve body: The part of a valve containing the passages for fluid flow, valve seat, and inlet and outlet connections.

Valve stem: The part of a valve that raises, lowers, or turns the valve disc.

Valve trim: The combination of the valve plug and the valve seat.

Wedge: The disc in a gate valve.

Wire drawing: The erosion of a valve seat under high velocity flow through which thin, wire-like gullies are eroded away.

Yoke bushing: The bearing between the valve stem and the valve yoke.

Additional Resources

This module presents thorough resources for task training. The following resource material is suggested further study.

Choosing the Right Valve. New York, NY: Crane Company.

Piping Pointers; Application and Maintenance of Valves and Piping Equipment. New York, NY: Crane Company.

Pipeline Mechanical. National Center for Construction Education and Research. Upper Saddle River, NJ: Prentice Hall.

The Piping Guide, 1980. San Francisco, CA: Syentek Books Company, Ltd.

Figure Credits

NCCER CURRICULA — USER UPDATE

NCCER makes every effort to keep its textbooks up-to-date and free of technical errors. We appreciate your help in this process. If you find an error, a typographical mistake, or an inaccuracy in NCCER's curricula, please fill out this form (or a photocopy), or complete the online form at **www.nccer.org/olf**. Be sure to include the exact module ID number, page number, a detailed description, and your recommended correction. Your input will be brought to the attention of the Authoring Team. Thank you for your assistance.

Instructors – If you have an idea for improving this textbook, or have found that additional materials were necessary to teach this module effectively, please let us know so that we may present your suggestions to the Authoring Team.

NCCER Product Development and Revision

13614 Progress Blvd., Alachua, FL 32615

Email: curriculum@nccer.org
Online: www.nccer.org/olf

❑ Trainee Guide   ❑ Lesson Plans   ❑ Exam   ❑ PowerPoints   Other _____

Craft / Level: _____   Copyright Date: _____

Module ID Number / Title: _____

Section Number(s): _____

Description: _____

_____

_____

_____

_____

Recommended Correction: _____

_____

_____

_____

_____

Your Name: _____

Address: _____

_____

Email: _____   Phone: _____

85208-13

Drawings and Detail Sheets

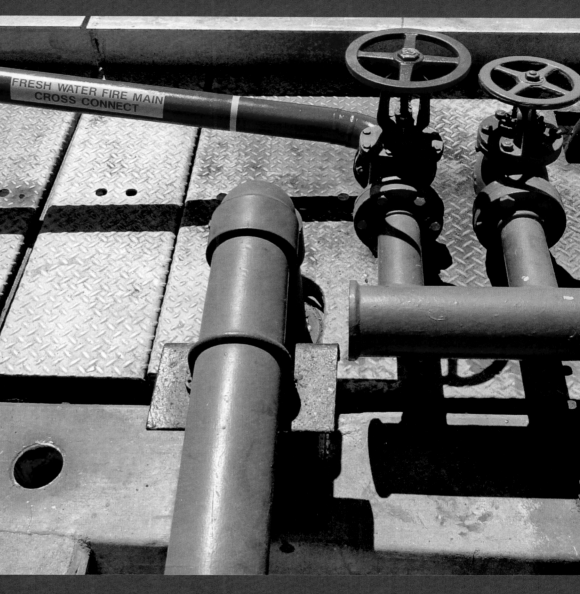

FRESH WATER FIRE MAIN
CROSS CONNECT

Module Eight

Trainees with successful module completions may be eligible for credentialing through NCCER's National Registry. To learn more, go to **www.nccer.org** or contact us at **1.888.622.3720**. Our website has information on the latest product releases and training, as well as online versions of our *Cornerstone* magazine and Pearson's product catalog.

Your feedback is welcome. You may email your comments to **curriculum@nccer.org**, send general comments and inquiries to **info@nccer.org**, or fill in the User Update form at the back of this module.

This information is general in nature and intended for training purposes only. Actual performance of activities described in this manual requires compliance with all applicable operating, service, maintenance, and safety procedures under the direction of qualified personnel. References in this manual to patented or proprietary devices do not constitute a recommendation of their use.

Objectives

When you have completed this module, you will be able to do the following:

1. Identify parts of drawings.
2. Identify types of drawings.
3. Identify drawings used by maritime pipefitters.

Performance Tasks

Under the supervision of the instructor, you should be able to do the following:

1. Sketch basic orthographic and isometric piping sections.
2. Identify types of drawings and parts of a drawing.

Trade Terms

Base line
Bill of materials
Buttock line (BTK)
Center line
Elevation view
Frame Line
Freeboard
General arrangement plan
Graphic scale
Isometric drawing (ISO)
Legend
Line number
Load water line (LWL)

Longitudinal frame line
Longitudinal frames
Midship line
Mold line (ML)
Orthographic projection
Piping and instrumentation drawing (P&ID)
Plan view
Section view
Sketch
Spool
Transverse frame line
Transverse frame
Water line

Industry-Recognized Credentials

If you are training through an NCCER-accredited sponsor, you may be eligible for credentials from NCCER's Registry. The ID number for this module is 85208-13. Note that this module may have been used in other NCCER curricula and may apply to other level completions. Contact NCCER's Registry at 888.622.3720 or go to **www.nccer.org** for more information.

Contents

Topics to be presented in this module include:

1.0.0 Introduction 1
2.0.0 Common Drawing Elements 1
  2.1.0 Title Blocks.. 1
  2.2.0 Scales and Measurements 2
  2.3.0 Lines 3
  2.4.0 Symbol and Abbreviation Legends.. 3
    2.4.1 Symbols 3
    2.4.2 Abbreviations 3
  2.5.0 Notes 4
  2.6.0 Revision Block 4
  2.7.0 Bill of Materials.. 4
3.0.0 Reference Lines 5
  3.1.0 Compartment Identification 6
4.0.0 Orthographic Projections 8
  4.1.0 Plan Views... 9
  4.2.0 Elevation Views 9
  4.3.0 Section Views 10
5.0.0 Types of Drawings 10
  5.1.0 General Arrangement Plan... 10
  5.2.0 Equipment Arrangement Drawings.. 17
  5.3.0 Piping and Instrumentation Drawings 17
  5.4.0 Isometric Drawings 17
  5.5.0 Spool Drawings 17
  5.6.0 Equipment (Vendor) Drawings... 17
  5.7.0 Pipe Support Drawings and Detail Sheets 21
6.0.0 Making Field Sketches 21
  6.1.0 Making Orthographic Sketches 21
  6.2.0 Isometric Sketches.. 24
    6.2.1 Making Isometric Piping Sketches 24
Appendix A Pipefitting Abbreviations.. 30
Appendix B Piping Symbols... 33
Appendix C Valve and Special Symbols... 35

Figures and Tables ───────────────────

Figure 1   Example of a title block..2
Figure 2   Graphic scale 3
Figure 3   Lines commonly used on drawings.............................4–5
Figure 4   Example of general notes..5
Figure 5   Revision block...6
Figure 6   Bill of materials..7
Figure 7   Reference lines in relation to the hull (2 sheets)......8–9
Figure 8   Orthographic views of an object................................. 10
Figure 9   Plan view...11
Figure 10 Ship elevation view... 12
Figure 11 Example of a section view.. 13
Figure 12 Example of a general arrangement plan (3 sheets)............... 14–16
Figure 13 Example of a piping and instrumentation drawing..................... 18
Figure 14 Simplified piping isometric drawing 19
Figure 15 Spool drawing 20
Figure 16 Symbols for pipe hangers and supports 22
Figure 17 Detail sheet for 5HR 23
Figure 18 Three-view orthographic sketch.................................... 24
Figure 19 Isometric base lines 25
Figure 20 Marking length, width, and height................................ 26
Figure 21 Completing sketch 25
Figure 22 Sketching dimensions 25

Table 1   Engineering Drawing Sizes.. 2

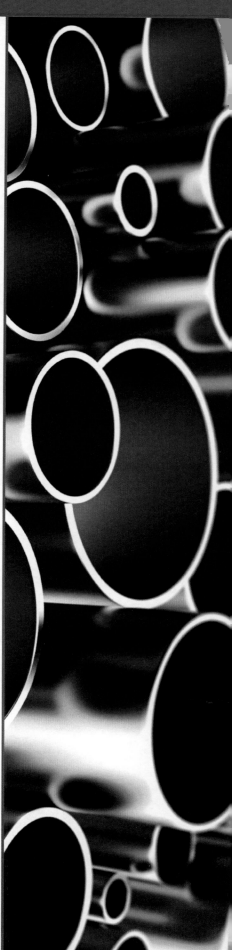

1.0.0 INTRODUCTION

When you are given a job to do, you will be given drawings that define how the job is to be performed and the materials that are to be used. If you are unable to correctly interpret the drawing, you will either be unable to do the job or might do it incorrectly. The ability to interpret drawings is therefore a key factor in your development as a craft professional.

Drawings are a graphic method of representing equipment, systems, and structures. A complete set of drawings for a vessel defines in detail everything from the complete assembly to the smallest piece-part.

> **NOTE**
> Drawings are often called blueprints because, at one time, the ammonia process used to copy drawings turned the paper blue.

A drawing is a detailed plan of a part, assembly, subassembly, or system. A detail sheet is a drawing, at a larger scale, of part of another drawing. Drawings and detail sheets are used by designers to provide the information needed to build or install equipment. The pipefitter uses drawings to fabricate piping runs, install them correctly, and to diagnose problems in piping systems.

> **NOTE**
> It is important to remember that workers must never deviate from a drawing. If something on the drawing appears incorrect, check with a supervisor before making a change.

2.0.0 COMMON DRAWING ELEMENTS

Certain elements are common to most drawings. These elements include the following:

- Title block
- Scales and measurements
- Symbol and abbreviation legends
- Notes
- Revision blocks
- Bill of materials

Other elements that may be found on drawings include drawing indexes and line lists. The drawing index, provided on the first page of the drawing set, is a complete list of all drawings included in the set. It is developed from the general arrangement plan, which is usually the first drawing created. The pages within each category are designated by a letter followed by a number. The drawings are usually organized into the following major categories, with the drawing category in parenthesis:

- Interior/graphics (D)
- Structural (S)
- Mechanical equipment (M)
- Piping and plumbing (P)
- Electrical (E)

The line list, or line table, is found on piping drawings. It provides the following information about piping runs:

- Line numbers in numerical order
- Class or specification of each line and size
- Whether or not each line is insulated
- What each line transports (liquid or gas)
- Starting and termination points
- Design pressure and temperature
- Operating pressure and temperature
- Testing pressure

Line numbering and other line information varies from one contractor to another. Some lines change as they change size or other characteristics, while others keep the same number from their beginning to their termination point. The piping supervisor usually fills out the line list and selects the piping specifications that can be used with the system.

2.1.0 Title Blocks

The title block (*Figure 1*) contains information used to identify the project or any part of the project. Title blocks are almost always located at the bottom right-hand corner of a drawing. The title block identifies the equipment represented on the drawing and the project to which it applies. The following key information is found in the title block:

- *Drawing title* – This explains what is drawn on that sheet. This may be the type of work, the work locations, or the name of the assembly or part covered by the drawing.
- *Sheet size* – There are standard sheet sizes (see *Table 1*).
- *Scale* – This shows the relationship between the size of the drawing and the size of the actual object. If the drawing is the same size as the object, the scale is full size.
- *Project number* – This is the code number assigned to the project.
- *Sheet number* – This indicates the sequence of this sheet in the series of drawings.
- *Revision number (Rev.)* – This indicates the current drawing. The number in this block can be

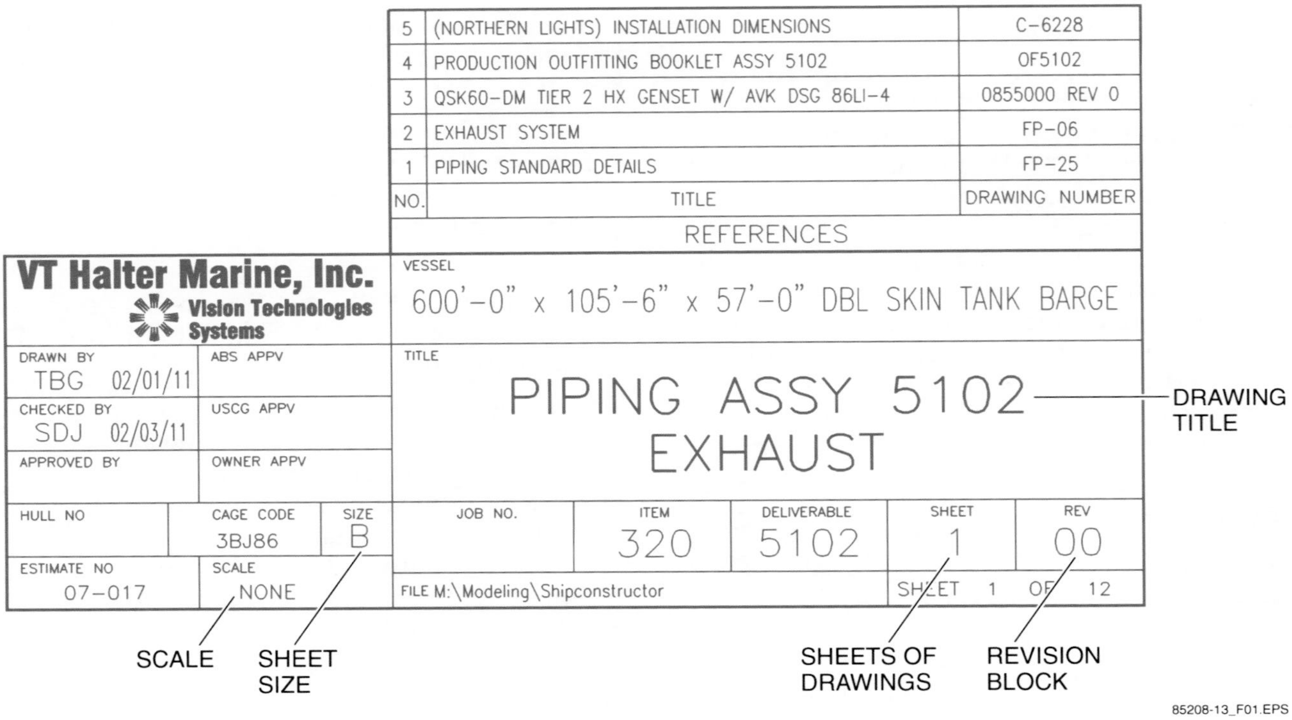

5	(NORTHERN LIGHTS) INSTALLATION DIMENSIONS	C−6228
4	PRODUCTION OUTFITTING BOOKLET ASSY 5102	OF5102
3	QSK60−DM TIER 2 HX GENSET W/ AVK DSG 86LI−4	0855000 REV 0
2	EXHAUST SYSTEM	FP−06
1	PIPING STANDARD DETAILS	FP−25
NO.	TITLE	DRAWING NUMBER

REFERENCES

VT Halter Marine, Inc.
Vision Technologies Systems

VESSEL
600'−0" x 105'−6" x 57'−0" DBL SKIN TANK BARGE

DRAWN BY TBG 02/01/11	ABS APPV
CHECKED BY SDJ 02/03/11	USCG APPV
APPROVED BY	OWNER APPV

TITLE

PIPING ASSY 5102 — DRAWING TITLE
EXHAUST

| HULL NO | CAGE CODE 3BJ86 | SIZE B |
| ESTIMATE NO 07−017 | SCALE NONE | |

| JOB NO. | ITEM 320 | DELIVERABLE 5102 | SHEET 1 | REV 00 |
| FILE M:\Modeling\Shipconstructor | | | SHEET 1 OF 12 | |

SCALE   SHEET SIZE   SHEETS OF DRAWINGS   REVISION BLOCK

85208-13_F01.EPS

Figure 1 Example of a title block.

referenced to the revision block for information about each revision.

- *Drawn by/Checked by/Approved by* – Lists the names or initials of the persons responsible for preparation of the drawing.
- *Zone* – Drawing sheets are zoned like road maps. Alpha characters start at the bottom and go up; numerical characters start at the right and go to the left. The title block in *Figure 1* is located in zone A1.

A key plan is provided if one plan drawing does not include the entire system being designed. The key plan shows where this drawing fits into the overall location. The shaded area shows the location of the drawing. The key plan may be shown on the **plan view** for reference.

2.2.0 Scales and Measurements

Scales on drawings using the US or customary system are usually written in inches or in feet and inches. Measurements of less than 1 foot are written as 0 feet and the number of inches. The scale of a drawing is written in the title block. It shows the size of the drawing as compared to the actual size of the object represented.

The scale of a drawing varies according to the size of the object being drawn and the size of the paper used. The drawing scale is usually indicated by a ratio between the size of the object and the size of the drawing. For example, a scale of 1 inch = 1 foot means that for every inch shown on the drawing, the actual size of the object is 1 foot. The standard scale of piping drawings is ⅜ inch = 1 foot.

Some drawings use the metric system. Since the metric system is based on units of ten, metric drawing scales are usually based on ratios. For instance, a 1:10 ratio means that 100 millimeters on the drawing equals 1,0000 millimeters (or 1 meter) on the ship. If needed, conversions can be worked out using readily available conversion charts. The letters U.N. stand for "unless noted otherwise on draw-

Table 1 Engineering Drawing Sizes

Sheet Size	Width (in)	Length (in)
A Horizontal	8.5	11.0
A Vertical	11.0	8.5
B	11.0	17.0
C	17.0	22.0
D	22.0	34.0
E	34.0	44.0
F	28.0	40.0
G	11.0	22.5 Min 90 Max
H	28.0	44.0 Min 143.0 Max
J	34.0	55.0 Min 176.0 Max
K	40.0	55.0 Min 143.0 Max

85208-13_T01.EPS

NCCER – *Maritime Pipefitting Level Two*   85208-13

ing." This indicates that some parts of the drawing may use a different scale.

Graphic scales on drawings (*Figure 2*) show distances that correspond with a unit of length used on the object in the drawing. They are usually placed in or near the title block of the drawing. When a drawing has not been held to a particular scale, and cannot usefully be measured, the scale block will have the initials N.T.S., for "not to scale."

2.3.0 Lines

Drawings are made up of lines. There are both solid and broken lines, as well as thick and thin lines. Each type has a specific meaning. Lines can define the actual physical dimensions and shapes of objects such as pipes, or can be used with arrows to indicate callouts, instructions, and notes. Most shipyard drawings use a mixture of standard and piping-specific lines. *Figure 3* shows some typical lines used in preparing drawings used by shipyard workers. Line usage may vary from one company to another, so it is important to understand the convention used in your company.

- Object (visible) lines are thick, continuous lines used to represent the outline of an object.
- Hidden (invisible) lines are thin lines made up of short dashes. They are used to represent features that would not be visible from the perspective in which the object is being viewed.
- Center lines consist of alternately long and short thin lines. They mark the center of an object. In the case of circles, both horizontal and vertical center lines are used. The intersection of the lines marks the exact center of the circle.
- Cutting plane lines are thick lines used to identify an alternate view of the area defined by the lines. Cutting plane lines usually terminate in arrows and are labeled with letters or a letter and number. The arrows show the direction in which the view is taken.
- Phantom lines are thin dashed lines used to show an alternate position of an object. Phantom lines are also used to identify future locations of equipment or structures.
- Dimension lines are thin, solid lines used to indicate the distance between two points on the drawing. Dimension lines are bound by extension lines.

- Leader lines are thin lines used to indicate a point at which a note, number, or other reference applies. They sometimes end in an arrowhead or dot. A leader line with a number might mean that the object being pointed to can be found at that item number listing on a bill of materials.
- Break lines are commonly used to shorten a long object so that it takes up less space on the drawing. Another use is to create a break in an object so that an object beneath it can be viewed.

2.4.0 Symbol and Abbreviation Legends

Symbols and abbreviations are used to simplify the preparation of drawings and to help convey a lot of information in a relatively small area. Although there are industry standards for symbols and abbreviations, companies may use their own conventions. Drawings may contain legends that identify the symbols and abbreviations they have used.

2.4.1 Symbols

Symbols are used to show objects, such as valves, fittings, flanges, and connections that would require too much detail and too much time if drawn literally. These objects are represented on drawings by small, simple line drawings. Symbols vary from job to job. *Appendix B* shows commonly used piping symbols. *Appendix C* shows some common valve and fitting symbols.

2.4.2 Abbreviations

Shipyard workers, including pipefitters, use industry-specific terms and phrases. To avoid having to write out each word or phrase every time it is used, standard abbreviations have been developed. The pipefitter should recognize these abbreviations and the meaning of each. *Appendix A* provides a list of commonly used pipefitting abbreviations. Note that there may be some abbreviations specific to a particular site. When in doubt, the pipefitter should consult someone who is more familiar with the job abbreviations.

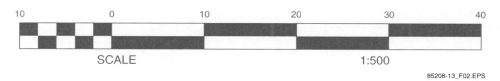

85208-13_F02.EPS

Figure 2 Graphic scale.

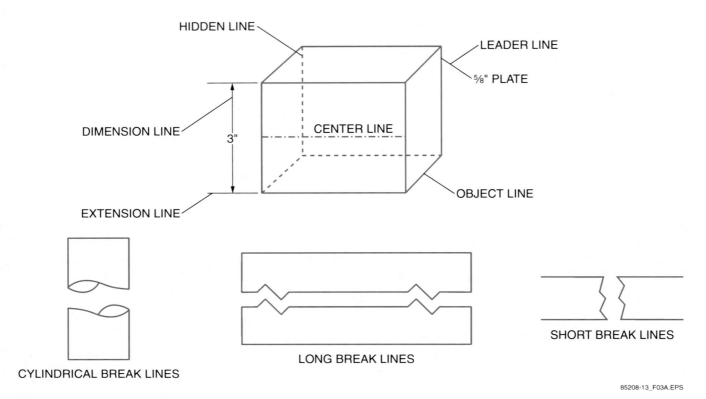

HIDDEN LINE

LEADER LINE

⅝" PLATE

DIMENSION LINE

CENTER LINE

3"

OBJECT LINE

EXTENSION LINE

CYLINDRICAL BREAK LINES

LONG BREAK LINES

SHORT BREAK LINES

85208-13_F03A.EPS

Figure 3 Lines commonly used on drawings. (1 of 2)

2.5.0 Notes

Notes are provided on drawings to provide information that cannot be easily represented in graphic form. Notes may be used to emphasize certain factors or to define specifications for materials or fabrication processes. Such notes may include general requirements, quality assurance criteria, calibration tolerances, and similar information. Notes may be located in a corner of a drawing, or they may be listed on a separate sheet and included with a set of drawings. Notes can be general or specific. General notes (*Figure 4*) can appear anywhere on a drawing, but usually appear at the upper left. In some cases, an entire sheet of the drawing may be dedicated to general notes.

General notes apply to the overall drawing. They may contain such information as material requirements, dimensions, requirements for hangers, or welding/brazing requirements. The main thing to recognize is that the general notes contain important information. The information in the general notes may not be found anywhere else on the drawing. Specific notes pertain to a specific location or item on the drawing and are generally tied to that location with a leader line.

When reading a drawing it is a good idea to first read the title block, and then the general notes.

WARNING!

Carefully read all notes before beginning any fabrication based on the drawing. Notes contain information that may directly affect personal protection and job quality.

2.6.0 Revision Block

The revision block is often located in the upper right hand corner of a drawing. It may also be located in the lower center of the drawing, or can be incorporated into the title block. The revision block is used to record changes that have been made to the drawing. All revisions are noted in this block, as shown in *Figure 5*. They are dated and are identified by a letter or number and a description of the revision. A drawing revision may be shown by the addition of a letter or number next to the original drawing number in the title block. It is also common to designate the locations where revisions occurred on the individual drawing sheets by placing a triangle containing the revision letter next to the place where a revision was made to the drawing.

2.7.0 Bill of Materials

The bill of materials, sometimes called the list of materials, is the key to identifying parts on a drawing. A bill of materials is used primarily on assembly drawings that show more than one part.

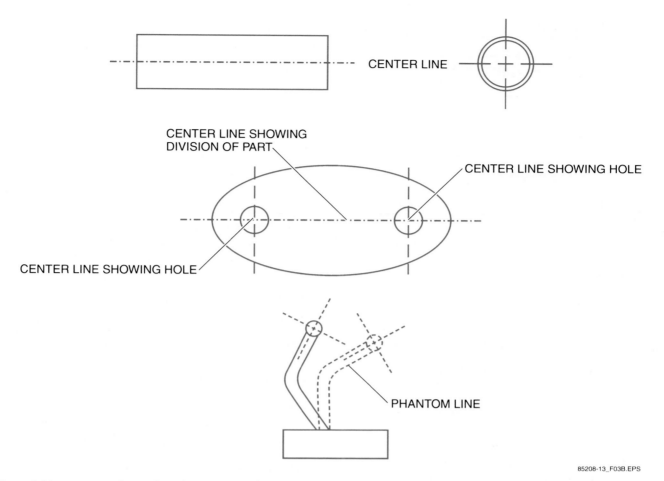

CENTER LINE

CENTER LINE SHOWING DIVISION OF PART

CENTER LINE SHOWING HOLE

CENTER LINE SHOWING HOLE

PHANTOM LINE

85208-13_F03B.EPS

Figure 3 Lines commonly used on drawings. (2 of 2)

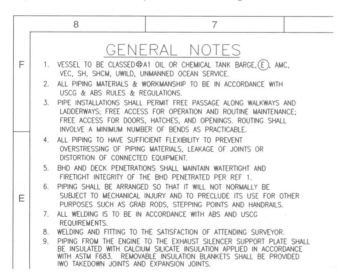

GENERAL NOTES

1. VESSEL TO BE CLASSED ⊕A1 OIL OR CHEMICAL TANK BARGE, Ⓔ, AMC, VEC, SH, SHCM, UWILD, UNMANNED OCEAN SERVICE.
2. ALL PIPING MATERIALS & WORKMANSHIP TO BE IN ACCORDANCE WITH USCG & ABS RULES & REGULATIONS.
3. PIPE INSTALLATIONS SHALL PERMIT FREE PASSAGE ALONG WALKWAYS AND LADDERWAYS; FREE ACCESS FOR OPERATION AND ROUTINE MAINTENANCE; FREE ACCESS FOR DOORS, HATCHES, AND OPENINGS. ROUTING SHALL INVOLVE A MINIMUM NUMBER OF BENDS AS PRACTICABLE.
4. ALL PIPING TO HAVE SUFFICIENT FLEXIBILITY TO PREVENT OVERSTRESSING OF PIPING MATERIALS, LEAKAGE OF JOINTS OR DISTORTION OF CONNECTED EQUIPMENT.
5. BHD AND DECK PENETRATIONS SHALL MAINTAIN WATERTIGHT AND FIRETIGHT INTEGRITY OF THE BHD PENETRATED PER REF 1.
6. PIPING SHALL BE ARRANGED SO THAT IT WILL NOT NORMALLY BE SUBJECT TO MECHANICAL INJURY AND TO PRECLUDE ITS USE FOR OTHER PURPOSES SUCH AS GRAB RODS, STEPPING POINTS AND HANDRAILS.
7. ALL WELDING IS TO BE IN ACCORDANCE WITH ABS AND USCG REQUIREMENTS.
8. WELDING AND FITTING TO THE SATISFACTION OF ATTENDING SURVEYOR.
9. PIPING FROM THE ENGINE TO THE EXHAUST SILENCER SUPPORT PLATE SHALL BE INSULATED WITH CALCIUM SILICATE INSULATION APPLIED IN ACCORDANCE WITH ASTM F683. REMOVABLE INSULATION BLANKETS SHALL BE PROVIDED IWO TAKEDOWN JOINTS AND EXPANSION JOINTS.

85208-13_F04.EPS

Figure 4 Example of general notes.

Figure 6 shows a very simple example of a bill of materials and its related drawing. Note the letters in the Item column. They relate to the four circled letters on the drawing. It is more common to use numbers, rather than letters, in the Item column.

The remaining parts of the bill of materials identify the parts by their part number and description and show the quantity required for each part. In general, only drawings for piece parts and small subassemblies will have a bill of materials on the same page as the image. For large drawing sets that cover hundreds of parts, the bill of materials may take several sheets of the drawing.

In addition to part number, description, and quantity, a bill of materials may contain references to the sheets of the drawing on which the parts appear. Part descriptions can range from one or two words to a very detailed description that includes the dimensions of the part.

3.0.0 REFERENCE LINES

One important function of drawings is to locate objects in relation to other objects or reference points. This information helps workers determine exactly where to install an object, drill a hole, or perform some other activity. Therefore, all measurements on drawings are stated in relation to something. For example, the location of a hole on a metal plate may be mea-

PROJ	NO	REVISION	RVSD	CHKD	APPD	DATE
3483	01	RELEASED FOR CONSTRUCTION		APD	NWS	JULY 06
3483	02	DELETED PART OF LINE 12037		APD	NWS	AUG 06
3483	03	⚠ ADDED WELDING SYMBOL		APD	NWS	AUG 06

85208-13_F05.EPS

Figure 5 Revision block.

sured from the edge of the plate. The location of a pipe opening in a bulkhead may be referenced to the center line of the ship. Common reference lines on a ship drawing are shown in *Figure 7*.

- The ship center line is a horizontal line running from bow to stern of the ship.
- The **base line** is a horizontal line at the bottom of the ship. This line is level and does not curve with the ship hull.
- The **midship line** is a vertical line halfway between the bow and stern. The midship line may be called the center line on some drawings.
- The **transverse frame line** represents frame members running across the ship from port to starboard at specified intervals. **Transverse frames** are often used as measurement reference points for distances and locations forward and aft. Locations of objects are often referenced in terms of their relationship to transverse frames, which are numbered.
- **Longitudinal frame lines** represent frame members running fore to aft along the length of the vessel. A **longitudinal frame** can be a shell, deck, or bulkhead stiffener running fore and aft.

> **NOTE**
> The terms *transverse* and *longitudinal* can be used in different ways. Sometimes they simply refer to an axis, while in other cases they are referring to specific structural members such as transverse frames.

Other reference lines are coordinated with one of the above-mentioned horizontal or vertical reference lines. The following are examples of commonly used reference lines:

- Buttock – A given distance from a center line.
- **Buttock line (BTK)** – A line running parallel to and at specific distances from the center line of the ship. These lines are used to measure widths when it is inconvenient to use the center line as a reference point.
- **Freeboard** – The distance between the water line and the deck or weather deck of a ship.
- **Load water line (LWL)** – The line at which a ship will float when loaded to its carrying capacity.
- **Mold line (ML)** – The exact location at which a structure is installed.
- **Water line** – A line parallel to the base line and at various distances depending on the displacement of the ship.

3.1.0 Compartment Identification

There are conventions for identifying compartments on a ship so that they can be located. This section describes the convention used to identify compartments on US Navy ships. The identification number assigned locates each compartment specifically, and generally indicates the function and use of the compartment. Compartment numbers consist of four parts, separated by hyphens, for example 5-165-0-E, in the following sequence:

Deck number
Frame number
Position in relation to center line of ship
Compartment use

Deck number – Numbering starts with the main deck, which is numbered 1. Lower decks are numbered consecutively 2 through n. The level above the main deck is numbered 01. These numbers continue consecutively, as well.

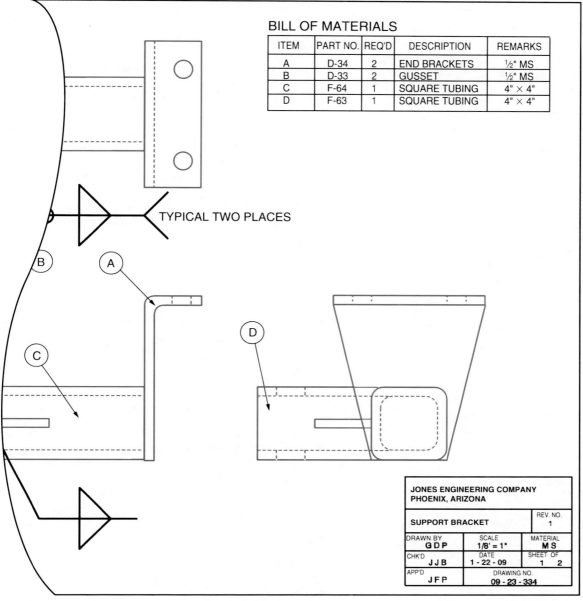

BILL OF MATERIALS

ITEM	PART NO.	REQ'D	DESCRIPTION	REMARKS
A	D-34	2	END BRACKETS	½" MS
B	D-33	2	GUSSET	½" MS
C	F-64	1	SQUARE TUBING	4" × 4"
D	F-63	1	SQUARE TUBING	4" × 4"

TYPICAL TWO PLACES

JONES ENGINEERING COMPANY PHOENIX, ARIZONA			
SUPPORT BRACKET			REV. NO. 1
DRAWN BY GDP	SCALE 1/8' = 1"		MATERIAL MS
CHK'D JJB	DATE 1 - 22 - 09		SHEET OF 1 2
APP'D JFP	DRAWING NO. 09 - 23 - 334		

85208-13_F06.EPS

Figure 6 Bill of materials.

> **NOTE**
>
> Decks above the main decks are referred to as levels.

Frame number – Transverse frames are consecutively numbered, based on frame spacing, until the aft transverse frame is reached. Forward of the forward perpendicular, frames are lettered starting from the perpendicular to the bull nose (A, B, C, etc.). Frames aft of the after perpendicular are double-lettered to the transom (AA, BB, CC, etc.). Compartments are numbered by the frame number of the foremost bulkhead of the compartment. If this bulkhead is located between frames, the number of the foremost frame within the compartment is used. Fractional numbers are not used except where frame spacing exceeds four feet.

Position in relation to ship center line – If a compartment is located so that the center line of the ship passes through it, a 0 is assigned to it. Compartments located completely to starboard of the center line are given odd numbers starting with 1, and those to port of center line are given even numbers starting with 2. If the center line of the ship passes through more than one compartment with the same forward bulkhead number, the compartment that has the portion of the for-

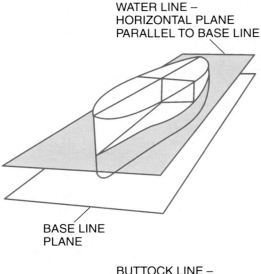

WATER LINE –
HORIZONTAL PLANE
PARALLEL TO BASE LINE

BASE LINE
PLANE

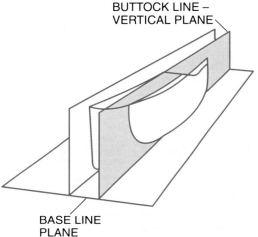

BUTTOCK LINE –
VERTICAL PLANE

BASE LINE
PLANE

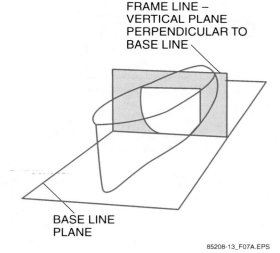

FRAME LINE –
VERTICAL PLANE
PERPENDICULAR TO
BASE LINE

BASE LINE
PLANE

85208-13_F07A.EPS

Figure 7 Reference lines in relation to the hull. (1 of 2)

ward bulkhead through which the center line of the ship passes is assigned the number 0 and the other carry numbers 01, 02, 03, etc.

Use of the compartment – A capital letter is used to identify the assigned primary use of the compartment. Only one capital letter is assigned, except that on dry and liquid cargo ships, a double letter identification is used to designate compartments assigned to carry cargo. Examples of compartment use are storage areas, various tanks, and living quarters.

Examples of these identifiers are the following:

A – Storage area
C – Ship and fire control operating spaces normally manned
E – Machinery (engineering) spaces that are normally manned
F – Fuel or fuel oil tanks
J – JP-5 tank
L – Living quarters
M – Ammunition stowage and handling
Q – Areas not otherwise covered
T – Vertical access trunk
V – Void

Recalling the compartment identifier 5 - 165 - 0 – E, it can be determined that the following is true:

- It is located four decks below the main deck.
- The foremost bulkhead is at frame 165.
- It is centered on the center line of the ship.
- It is used as an engineering (E) space.

4.0.0 ORTHOGRAPHIC PROJECTIONS

Most ship assembly drawings use an illustration technique known as **orthographic projection**. An orthographic projection is a drawing that shows the shape of a three-dimensional object by using a series of two-dimensional views. It is made by extending perpendicular (90-degree) lines from an object to create a new projected view.

Orthographic drawings illustrate the exact shape of an object from different viewing points. An orthographic projection is made by extending lines from the surfaces and features of an object in one view to create a new, projected view. Each view shows the object as it is seen from a certain direction or point of view. By combining the views, a drafter can provide all the information needed to fabricate the object.

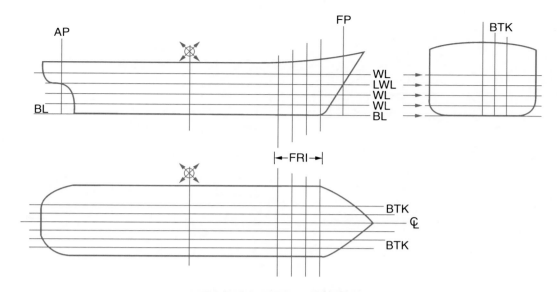

ABBREVIATIONS LEGEND	
BASE LINE	BL
WATERLINE	WL
CENTER LINE	℄
BUTTOCK LINE	BTK
FRAME LINE	FR
MIDSHIP LINE	⊗
LOAD WATER LINE	LWL
FORWARD PERPENDICULAR	FP
AFTER PERPENDICULAR	AP

85208-13_F07B.EPS

Figure 7 Reference lines in relation to the hull. (2 of 2)

Orthographic views are ideally drawn at flat viewing angles that do not create distortion from perspective. A front view, for example, is typically a head-on view of a major part surface. It shows features (both visible and hidden) of the part as they would appear from the front of the object. However, to show the width/depth of the object, a second view from the top or the side is required. Combined, the multiple views give fabricators a complete visual representation of the object.

Even with multiple views, complex or oddly shaped parts can be difficult to visualize. When comparing an object's three common views (front, side, and top), it can be helpful to think of them as making a box. The box represents the object in three dimensions. When unfolded and made flat, the three views are brought into the plane of the paper (*Figure 8*). This is how the orthographic projection is arranged.

Orthographic drawings can be grouped into three types, depending on the imaginary point from which they are viewed.

4.1.0 Plan Views

A plan view is a drawing of an object or area from above it, looking down on the part or system from the top. *Figure 9* is an example of a plan view. A plan view gives two dimensions, length and width. Its primary reference lines are the **frame lines** and center line. Transverse section lines are references for **section views** and longitudinal section lines are used for **elevation views.**

The plan view shows the exact locations of the equipment in relation to established reference points, objects, or surfaces. These reference points may be bulkheads, reference lines, or other equipment. The plan view is the view from which the elevation and section views are derived.

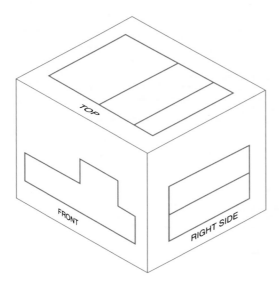

Figure 8 Orthographic views of an object.

4.2.0 Elevation Views

The definition of elevation and section drawings will vary, depending on whether the subject is a ship or another type of vessel. For ships, an elevation view is a side view looking from port to starboard. An elevation view gives the height and length of an object. Its primary references are the frame lines, molded lines of the deck, and the base line of the ship. Elevations are usually drawn looking toward the top of the left side of the drawing. Sometimes elevation drawings are called front, rear, or side elevations. *Figure 10* is an example of an elevation drawing. It represents the view defined by cutting plane lines 6-B on the plan view of *Figure 9*.

For structures other than ships, an elevation view looks straight at the structure from any side. The front elevation of a drill rig, for example, would look at the front of the rig from top to bottom.

4.3.0 Section Views

It often happens that a plan or elevation view is not able to show all the necessary information. In such cases, a section view is used to show internal features of an assembly or structure. Keep in mind, however, that section views for ships are different from those of other types of vessels and structures. In ship drawings, a section view is always a view of an object looking fore or aft. A section view shows the width and height dimensions. Its primary reference lines are the molded lines of the decks, the ship center line, and the ship base line. *Figure 11* is an example of such a section view. It represents the view defined by cutting plane lines 8A-B on the plan view of *Figure 9* and is a view of the stack looking forward. It should be noted that cutting plane lines are not always used to designate sections for ship drawings because a section can be any view from a fore and aft perspective.

In drawings for other types of vessels and structures, section views can be taken from any perspective. They are generally vertical, but horizontal or plan sections may be included. In section views, the parts are usually labeled. A section view shows the width and height dimensions.

5.0.0 TYPES OF DRAWINGS

Most drawings used by pipefitters are plan, elevation, or section views. Some views, however, are specific enough to have their own classifications. Pipefitters use the following types of drawings during the ship-building process:

* General arrangement plan
* Equipment arrangement drawings
* **Piping and instrumentation drawings (P&ID)**
* **Isometric drawings (ISO)**
* **Spool** drawings
* Equipment or vendor drawings
* Pipe support drawings and detail sheets
* As-built drawings

5.1.0 General Arrangement Plan

The general arrangement (GA) plan (*Figure 12*) is the first, or top, drawing on a set of drawings. A general arrangement plan is usually a plan view, and shows the arrangement of major subsections of the complete unit to be constructed. The general arrangement plan is seldom used for actual construction, but serves as a guide for locating the de-

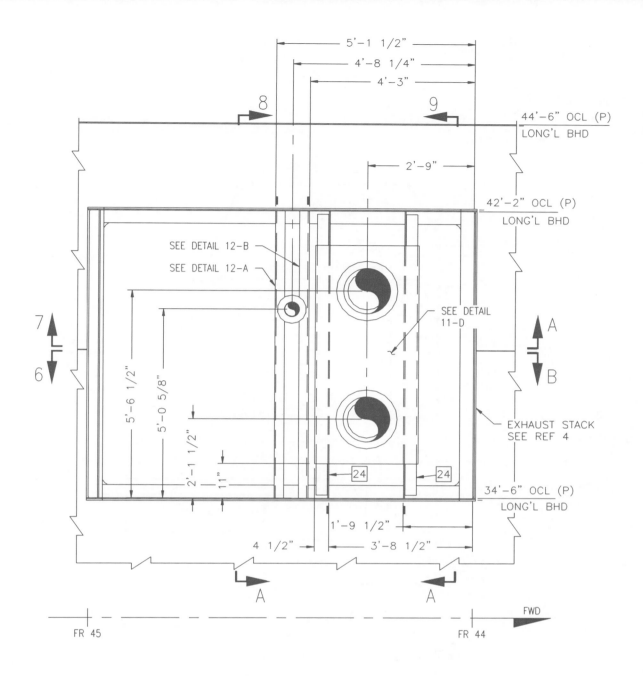

PLAN VIEW 6-A
EXHAUST STACK @ TOP OF HOUSE
FRAMES 44-45

HULL NO	CAGE CODE	SIZE	JOB NO.	ITEM	DELIVERABLE	SHEET	REV
	3BJ86	B		320	5101	6	00
ESTIMATE NO	SCALE		FILE			SHEET 6	
07-017	NONE						

85208-13_F09.EPS

Figure 9 Plan view.

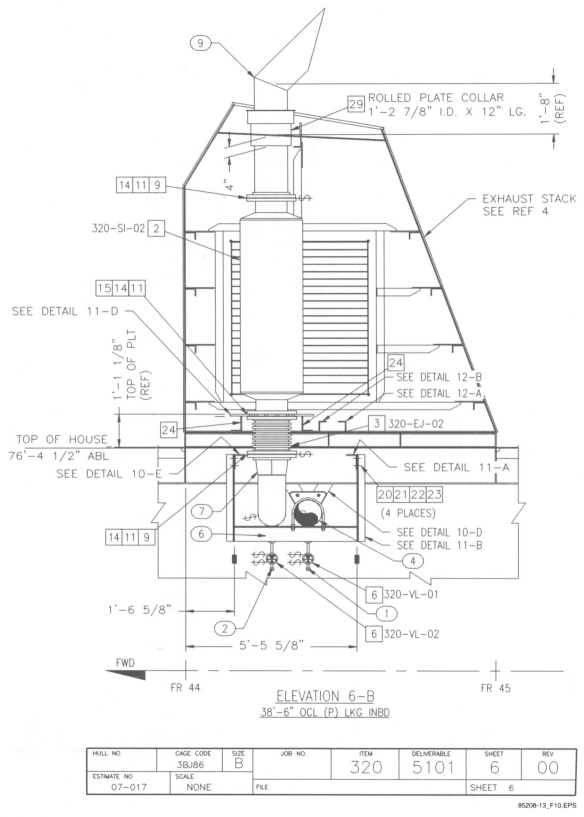

Figure 10 Ship elevation view.

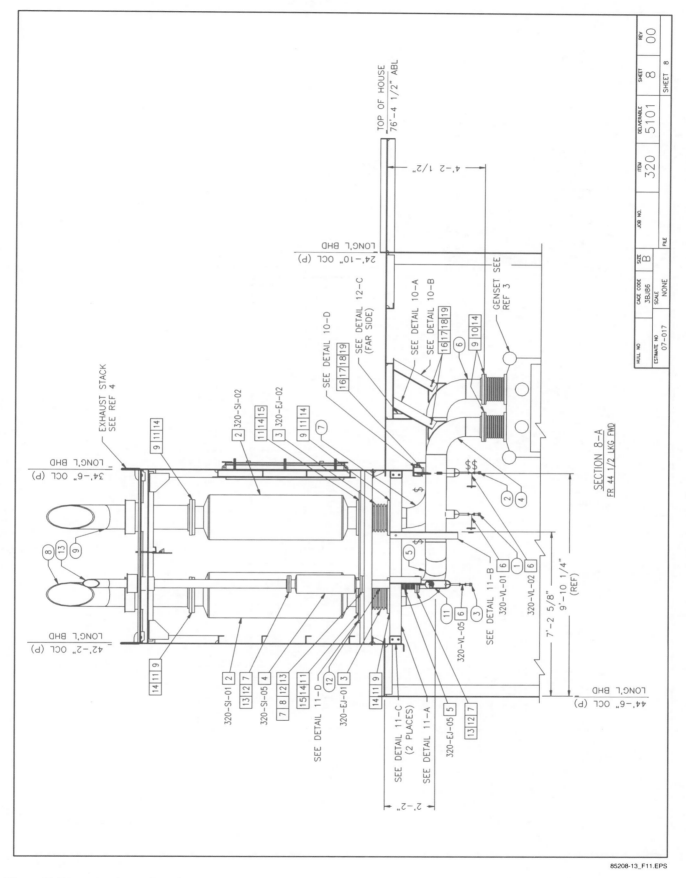

Figure 11 Example of a section view.

 85208-13 Drawings and Detail Sheets

Module Eight   13

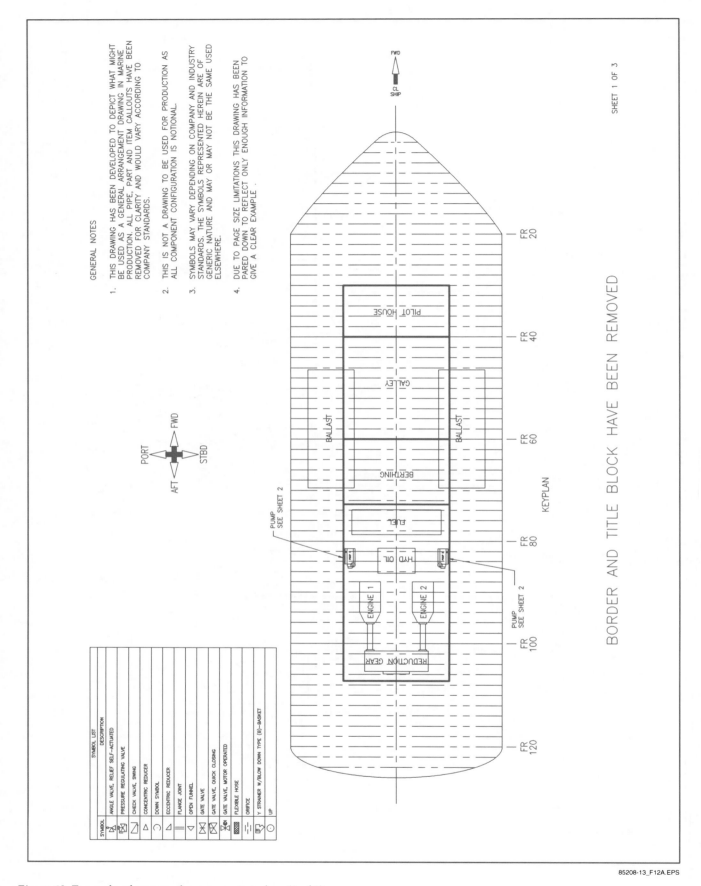

GENERAL NOTES

1. THIS DRAWING HAS BEEN DEVELOPED TO DEPICT WHAT MIGHT BE USED AS A GENERAL ARRANGEMENT DRAWING IN MARINE PRODUCTION. ALL PIPE, PART AND ITEM CALLOUTS HAVE BEEN REMOVED FOR CLARITY AND WOULD VARY ACCORDING TO COMPANY STANDARDS.

2. THIS IS NOT A DRAWING TO BE USED FOR PRODUCTION AS ALL COMPONENT CONFIGURATION IS NOTIONAL.

3. SYMBOLS MAY VARY DEPENDING ON COMPANY AND INDUSTRY STANDARDS. THE SYMBOLS REPRESENTED HEREIN ARE OF GENERIC NATURE AND MAY OR MAY NOT BE THE SAME USED ELSEWHERE.

4. DUE TO PAGE SIZE LIMITATIONS THIS DRAWING HAS BEEN PARED DOWN TO REFLECT ONLY ENOUGH INFORMATION TO GIVE A CLEAR EXAMPLE .

BORDER AND TITLE BLOCK HAVE BEEN REMOVED

SYMBOL LIST	
SYMBOL	DESCRIPTION
	ANGLE VALVE, RELIEF SELF–ACTUATED
	PRESSURE REGULATING VALVE
	CHECK VALVE, SWING
	CONCENTRIC REDUCER
	DOWN SYMBOL
	ECCENTRIC REDUCER
	FLANGE JOINT
	OPEN FUNNEL
	GATE VALVE
	GATE VALVE, QUICK CLOSING
	GATE VALVE, MOTOR OPERATED
	FLEXIBLE HOSE
	ORIFICE
	Y STRAINER W/BLOW DOWN TYPE (B)–BASKET
	UP

KEYPLAN

FWD
CL
SHIP

FR 20
FR 40
FR 60
FR 80
FR 100
FR 120

PILOT HOUSE
GALLEY
BALLAST
BALLAST
BERTHING
FUEL
HYD OIL
ENGINE 1
ENGINE 2
REDUCTION GEAR

PUMP SEE SHEET 2
PUMP SEE SHEET 2

PORT
STBD
AFT FWD

85208-13_F12A.EPS

Figure 12 Example of a general arrangement plan. (1 of 3)

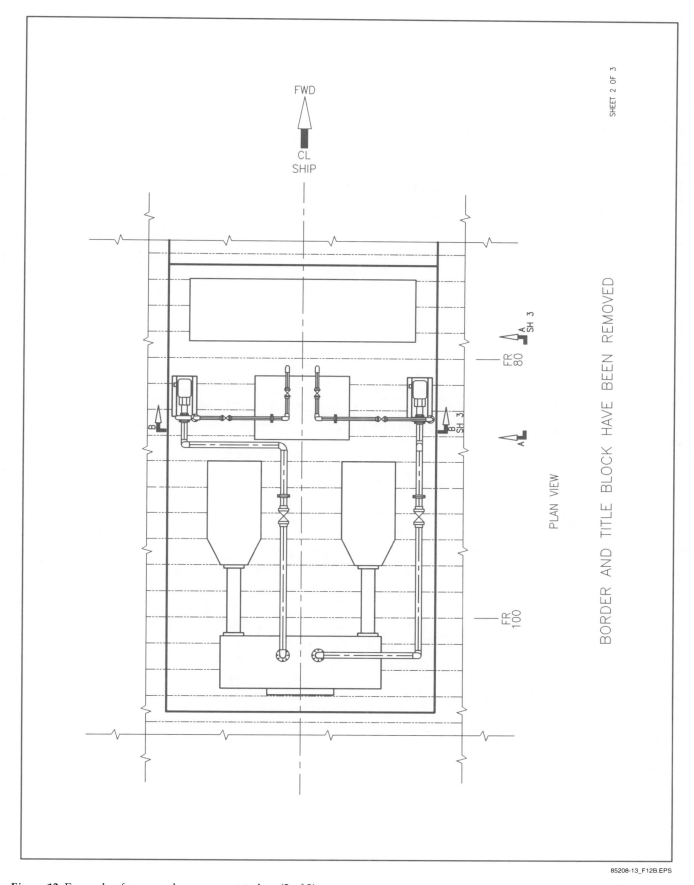

FWD

CL
SHIP

FR 80

A
SH 3

B

A
SH 3

A

B

FR 100

PLAN VIEW

BORDER AND TITLE BLOCK HAVE BEEN REMOVED

85208-13_F12B.EPS

Figure 12 Example of a general arrangement plan. (2 of 3)

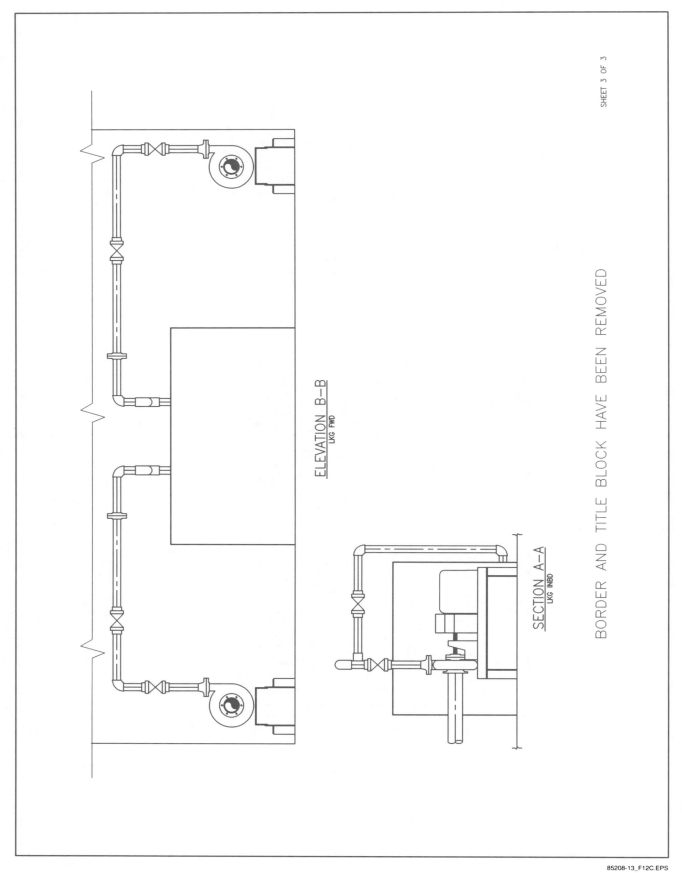

ELEVATION B–B
LKG FWD

SECTION A–A
LKG INBD

BORDER AND TITLE BLOCK HAVE BEEN REMOVED

85208-13_F12C.EPS

Figure 12 Example of a general arrangement plan. (3 of 3)

tailed drawings used for construction. For piping systems, a GA plan is an orthographic drawing of all the piping and related systems in a given space. It is often divided into drawings of individual systems. For example, a compartment may have a GA drawing showing only the fresh water piping.

5.2.0 Equipment Arrangement Drawings

Equipment arrangement drawings show how the equipment is arranged and the locations of all equipment used on the ship. Equipment on these drawings is usually drawn to correspond to the basic shape of the item represented. Detail may be eliminated when exact drawings exist elsewhere. The entire drawing is to scale and shows the entire piece of equipment. Equipment arrangement drawings are more detailed than GA drawings.

5.3.0 Piping and Instrumentation Drawings

Piping and instrumentation drawings (P&IDs) are schematic diagrams of a complete piping system or systems (*Figure 13*). P&IDs show the process flow, and may be called process flow diagrams. They also show all equipment, pipelines, valves, instruments, and controls necessary for the operation of that system. Because their purpose is only to provide a representation of the work to be done, P&IDs are not drawn to scale or dimensioned. Pieces of equipment on P&IDs are usually drawn to correspond with the actual shape of the object. Valves and other parts are usually indicated with symbols. P&IDs often show an area that is enclosed by a dashed line and has the initials VS inside the enclosed area. The VS indicates that all of the equipment within the enclosed area is vendor-supplied; therefore, the pipefitters are only responsible for the piping up to that area.

5.4.0 Isometric Drawings

An isometric drawing, or ISO, combines the plan and elevation views of a system or part of a system into one drawing (*Figure 14*). Note that isometric drawings are a departure from the normal practice of using orthographic projections. The purpose of an isometric drawing is to show a three-dimensional representation of the finished assembly. Usually, only one piping run is shown on a single isometric drawing, although for simple or duplicate pipelines, more than one line may be shown. An isometric pipe drawing shows the details needed to fabricate and install a section of a piping system. An ISO also usually includes a bill of materials, which lists the types of pipe,

flanges, valves, bolts, and gaskets required for the pipe run.

Isometric piping drawings use the same symbols as piping schematics or any other piping drawings. The ISO has arrows that show the direction of flow. To read an ISO, begin at the end of the pipe run nearest the source and work in the direction of flow. Trace the line from the first fitting, and identify all the fittings and components by their symbols.

5.5.0 Spool Drawings

A spool is a prefabricated section of piping. A spool drawing (*Figure 15*) is a representation of a section of piping that is to be fabricated before erection. Spool drawings are usually drawn for each line in the system. They provide the detail dimensions of each line. A spool drawing usually includes the following information:

- Instructions the pipefitter and welder need to fabricate the spool
- Cut lengths of pipe
- Any fittings and flanges required to fabricate the spool
- Materials required, or specifications
- Any special treatment required
- How many spools of the same type are required

Bolts, gaskets, valves, and instruments are not usually shown on a spool drawing. Most spools are made in a shop rather than in the field. The size of an individual spool is usually limited by the ability to transport it to the location where it will be installed. Generally, a spool is limited in dimensions to about 40 feet by 10 feet by 8 feet (12 meters by 3 meters by 2.4 meters).

5.6.0 Equipment (Vendor) Drawings

Many outside companies, called vendors, supply shipbuilders with vital components, including some large pieces of equipment. These equipment manufacturers supply drawings of the equipment. The drawings include the fabrication, erection, and setting drawings that are necessary to install a piece of equipment. They may also include manufacturer's standard drawings or catalog sheets, performance charts, brochures, and other data. The vendor may also be required to provide samples.

Pipefitters use equipment vendor drawings to install equipment and to ensure that all the necessary pre-piping is in its required location. Equipment drawings are required for almost every product that is fabricated away from the shipbuilding site.

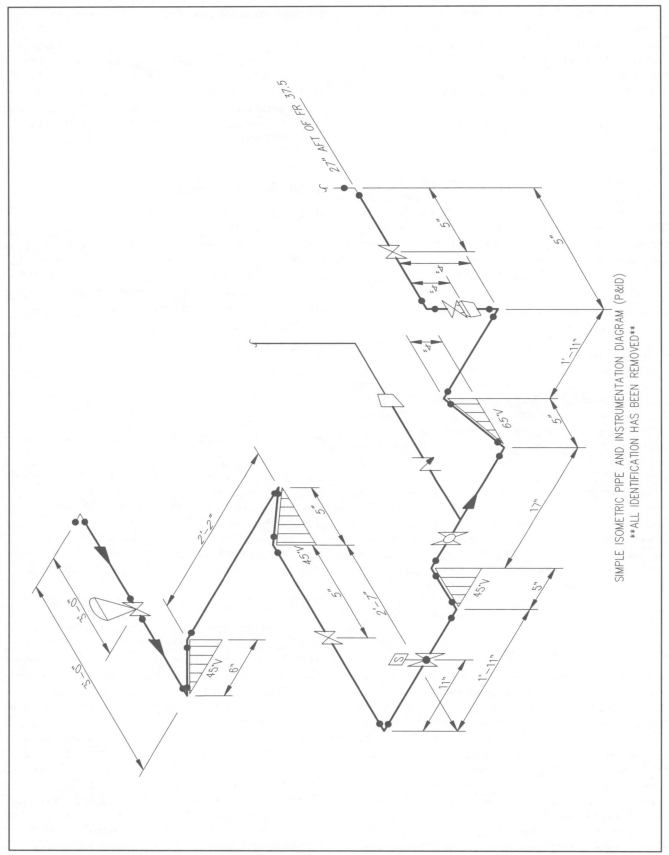

Figure 13 Example of a piping and instrumentation drawing.

85208-13_F13.EPS

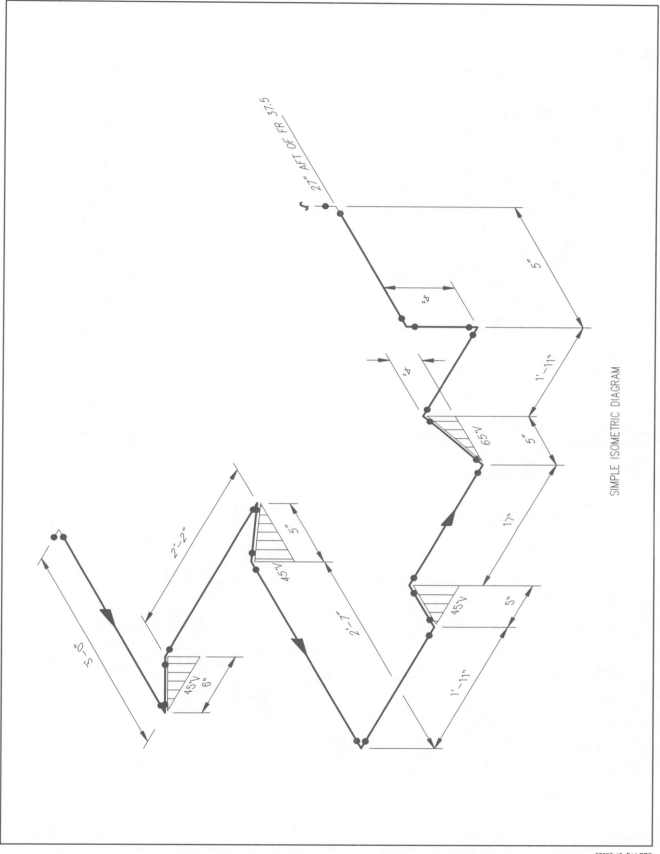

SIMPLE ISOMETRIC DIAGRAM

Figure 14 Simplified piping isometric drawing.

85208-13_F14.EPS

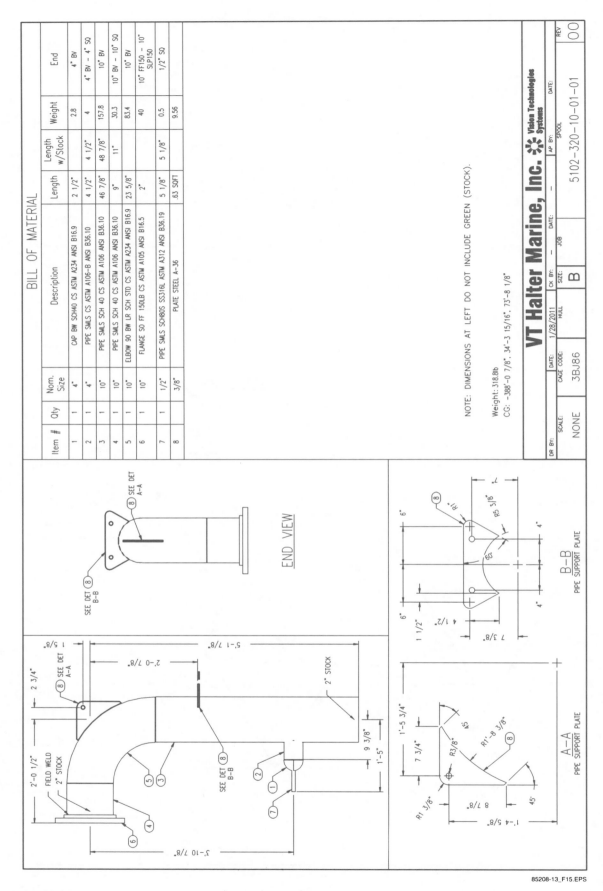

BILL OF MATERIAL

Item #	Qty	Nom. Size	Description	Length	Length w/Stock	Weight	End
1	1	4"	CAP BW SCH40 CS ASTM A234 ANSI B16.9	2 1/2"		2.8	4" BV
2	1	4"	PIPE SMLS CS ASTM A106-B ANSI B36.10	4 1/2"	4 1/2"	4	4" BV – 4" SQ
3	1	10"	PIPE SMLS SCH 40 CS ASTM A106 ANSI B36.10	46 7/8"	48 7/8"	157.8	10" BV
4	1	10"	PIPE SMLS SCH 40 CS ASTM A106 ANSI B36.10	9"	11"	30.3	10" BV – 10" SQ
5	1	10"	ELBOW 90 BW LR SCH STD CS ASTM A234 ANSI B16.9	23 5/8"		83.4	10" BV
6	1	10"	FLANGE SO FF 150LB CS ASTM A105 ANSI B16.5	2"		40	10" FF150 – 10" SLP150
7	1	1/2"	PIPE SMLS SCH80S SS316L ASTM A312 ANSI B36.19	5 1/8"	5 1/8"	0.5	1/2" SQ
8		3/8"	PLATE STEEL A-36	6.3 SQFT		9.56	

NOTE: DIMENSIONS AT LEFT DO NOT INCLUDE GREEN (STOCK).

Weight: 318.8lb
CG: -388'-0 7/8", 34'-3 15/16", 73'-8 1/8"

END VIEW

A–A
PIPE SUPPORT PLATE

B–B
PIPE SUPPORT PLATE

VT Halter Marine, Inc. Vision Technologies Systems

| DR BY: | SCALE: NONE | CAGE CODE: 3BJ86 | DATE: 1/28/2011 | JOB | DATE: | AP BY: | DATE: |
| | | | CK BY: | SIZE: B | HULL | SPOOL 5102-320-10-01-01 | REV 00 |

85208-13_F15.EPS

Figure 15 Spool drawing.

Equipment vendor drawings are usually prepared after the contract is awarded and construction is under way. They are prepared by the equipment vendors or fabricators and are approved by the primary contractor.

Vendor drawings are stamped either certified or noncertified by the manufacturer. If the drawings are stamped certified, the manufacturer guarantees that the piece of equipment is made exactly to the dimensions shown in the drawing. If the piece of equipment is different than the drawing shows, the manufacturer is held liable for the equipment. If the drawing is stamped noncertified, the equipment is not guaranteed to the specifications shown in the drawing. *Noncertified* drawings and equipment are much less expensive than certified drawings and equipment.

5.7.0 Pipe Support Drawings and Detail Sheets

Before a piping system is installed, engineers analyze the system and determine the type, size, and placement of all hangers and supports in the system. Proper pipe hangers and supports are vital to ensure that the pipes will not sag, vibrate, or break. Piping drawings are then made, showing the placement of each type of hanger and support. The pipefitter must be able to read and interpret these drawings to install the hangers and supports in the proper place.

Symbols for pipe supports vary from job to job. On piping drawings, the hanger or support has a reference number that refers to the specific detail sheets used on that job. Check the detail sheets to determine the exact type of hanger to use. Check the company's standards and specifications to determine the reference prefixes used on the job site.

Each engineering company also uses pipe hanger and support symbols that are specific to a job site. These symbols appear on plan and isometric drawings and are called out by the reference number as previously explained. *Figure 16* shows an example of pipe support drawings and symbols.

The detail sheets of a pipe hanger symbol are used to determine the details of a particular hanger. The detail sheets include all of the information needed to fabricate and install a hanger. Often a number of detail sheets are copied onto one page, at the back of a set of drawings, or added to the drawings on which the details are applicable. Notice the symbol in the upper left-hand corner of *Figure 16*. It designates a clevis hanger with reference number 5HR. *Figure 17* shows the detail sheet for 5HR.

6.0.0 MAKING FIELD SKETCHES

The overall design of a ship may not include all of the details of piping systems. Sometimes, the pipefitter must plan the exact layout of a piping system. By consulting the working drawings and making a **sketch** of the work to be done, pipefitters can check materials and identify potential difficulties on the project. This saves time and labor. Sketches are commonly made by pipefitters and then sent to shop services to aid them in fabricating a piping run.

Even when others have designed piping, it is helpful to make sketches of the parts and the general positioning of the pipes. Generally, the full set of plans is too large to carry to a specified point of installation. By making a detailed sketch, the pipefitter has a ready reference to take to the work site. Sketches save time and help avoid mistakes.

In addition to reading and interpreting drawings, pipefitters must make freehand sketches of parts or assembled units to convey information about how parts should be made or assembled. Sketching is a means of conveying an idea rather than a method of making complete, perfect drawings. Pipefitters must develop some basic sketching skills in order to produce sketches that are clear and easy to read.

To produce clear sketches, you must be able to draw lines, angles, arcs, ellipses, and circles. You must be able to make sketches that are in correct proportion to the object being sketched, including dimensions. As with any skill, field sketching takes considerable practice. As you learn these skills, practice the procedures so that you will continually do a better job. The following sections explain the basic skills needed to make orthographic and isometric sketches.

6.1.0 Making Orthographic Sketches

Orthographic sketches are the easiest freehand sketches to make. Orthographic sketches show only one side of an object at a time, but they usually show two or three views of the object. Drawing the top, front, and sides is done in a way that relates them to each other. The views are developed is the same way as in any drawing, and the same types of lines previously shown are used. The difference is that orthographic sketches are drawn freehand. Sketches of small parts are made as near the actual size as possible or to scale. *Figure 18* is an example of a three-view orthographic sketch.

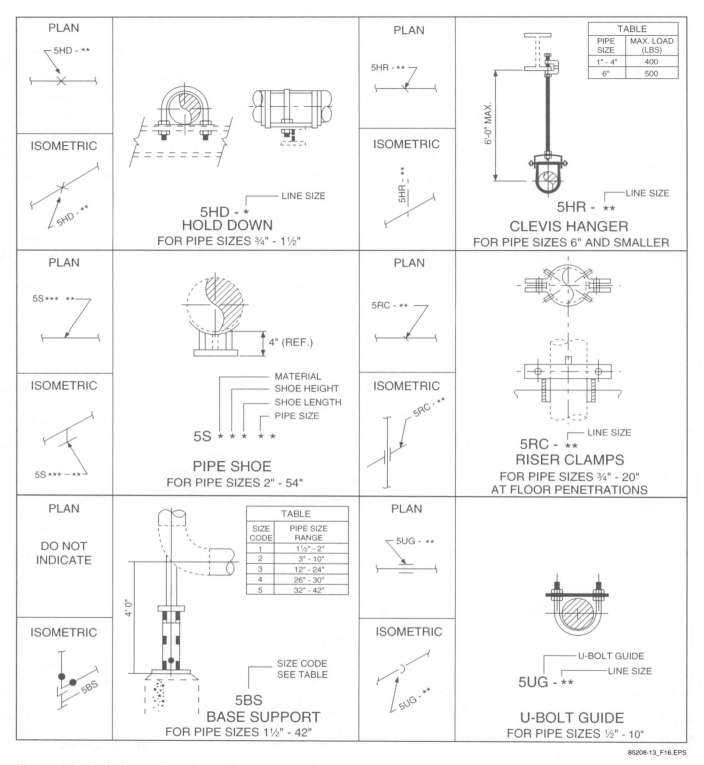

Figure 16 Symbols for pipe hangers and supports.

85208-13_F16.EPS

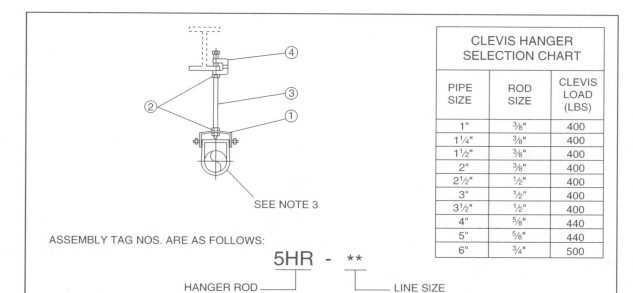

ASSEMBLY TAG NOS. ARE AS FOLLOWS:

5HR - ✳✳

HANGER ROD ⎯⎯ DESIGNATION

LINE SIZE

CLEVIS HANGER SELECTION CHART

PIPE SIZE	ROD SIZE	CLEVIS LOAD (LBS)
1"	3/8"	400
1 1/4"	3/8"	400
1 1/2"	3/8"	400
2"	3/8"	400
2 1/2"	1/2"	400
3"	1/2"	400
3 1/2"	1/2"	400
4"	5/8"	440
5"	5/8"	440
6"	3/4"	500

SEE NOTE 3

FABRICATION NOTES FOR METALLIC APPLICATIONS

1. PARTS EQUAL TO POWER PIPING COMPONENTS MAY BE USED.
2. FIELD TO CUT TO LENGTH REQUIRED AND TO THREAD BOTH ENDS.
3. FOR STAINLESS STEEL PIPE (1/2" - 6") USE 16-GAUGE STAINLESS STEEL BEARING PLATE.

TABLE 1			
LINE SIZE	CLAMP FIGURE NO.	BEAM BRACKET & BOLT SIZE	ROD DIA. & WELDLESS EYE & HEX NUT SIZE
1	222	1/2"	3/8"
2	222	1/2"	3/8"
3	222	1/2"	1/2"
4	222	1/2"	3/4"
6	222	3/4"	3/4"

MARK NO.	DESCRIPTION	TYPE	QUAN. REQ'D
①	ADJUSTABLE CLEVIS HANGER, POWER PIPING FIG. NO. 11 (SEE NOTE 1)	5HR	1
②	HEX NUT SAME SIZE AS ROD DIAMETER, POWER PIPING FIG. NO. 61 (SEE NOTE 1)	5HR	3
③	6'-0" LONG CARBON STEEL ROD, ASTM-A36, NOT THREADED, DIAMETER PER TABLE 1 (SEE NOTE 2)	5HR	1
④	CLAMP WITH RETAINING CLIP	5HR	1

ENGINEERING COMPANY NAME	HANGER RODS **5HR - ✳✳**		
	* = LINE SIZE		
	SPECIFICATION NUMBER		
	50201	SHEET	1 OF 1

85208-13_F17.EPS

Figure 17 Detail sheet for 5HR.

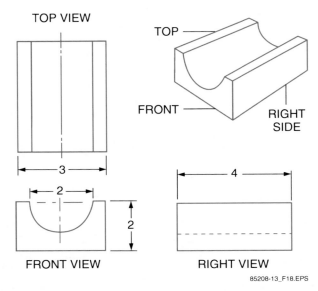

TOP VIEW

TOP

FRONT

RIGHT SIDE

3

2

2

FRONT VIEW

4

RIGHT VIEW

85208-13_F18.EPS

Figure 18 Three-view orthographic sketch.

Follow the steps below to make an orthographic sketch of a rectangular object. It is usually best to start by drawing the front, or longest, view.

Step 1 Imagine that you are looking directly at the front of the object.

Step 2 Draw the front of the object, as it would appear from the front.

Step 3 Imagine that you are looking directly at the left (your right) side of the object.

Step 4 Draw the left side of the object, as it would appear from the side.

Step 5 Imagine that you are looking directly down on the object.

Step 6 Draw the top of the object, as it would appear when looked down upon.

Step 7 Observe the three drawings. If they do not give a clear picture of the object, add views as necessary.

Step 8 Add height, width, and length dimensions to the drawing.

> **NOTE**
>
> Sometimes it is possible to completely illustrate an object with only two drawings. In these cases drawing the third view is not necessary.

6.2.0 Isometric Sketches

Isometric sketches show two or more sides of the object in a single view. An isometric sketch is developed around three major lines called isometric base lines or axes. The left-side and the right-side isometric base lines each form a 30-degree angle to the horizontal base line. The vertical base line forms a 90-degree angle to the horizontal base line. *Figure 19* shows the isometric base lines.

Follow these steps to make an isometric sketch of a rectangular object:

Step 1 Sketch the three isometric base lines.

Step 2 Mark the length of the object on the right-side isometric base line.

Step 3 Mark the width of the object on the left-side isometric base line.

Step 4 Mark the height of the object on the vertical isometric base line. *Figure 20* shows a sketch marked with the length, width, and height of the object.

Step 5 Sketch the remaining lines parallel to the three isometric base lines to complete the sketch as shown in *Figure 21*.

Step 6 Sketch the dimensions (*Figure 22*).

Step 7 Erase all unnecessary lines.

6.2.1 Making Isometric Piping Sketches

Isometric sketches are used to show how to fabricate and install a piping run. Follow the steps below to make an isometric sketch of a piping system:

Step 1 Study the orthographic view for enough time to completely understand the pipe section that it represents.

- Have a clear picture of the orientation of the pipe section with the rest of the project.
- Understand all dimensions, flow directions, elevations, line drops, and how it is to be put together.
- Note all needed components, such as valves and instrument flanges.

Step 2 If necessary, make some preliminary sketches on scrap paper. After making as many preliminary sketches as necessary, proceed to Step 3.

Step 3 Draw the pipe lines in proper relation to the orthographic drawing. Remember that the lines do not have to be drawn to scale.

Step 4 Add any component details as necessary. When applicable, use the proper standard symbols for these components.

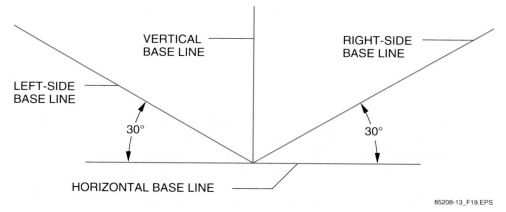

Figure 19 Isometric base lines.

85208-13_F19.EPS

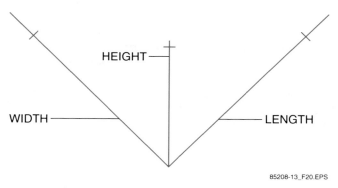

Figure 20 Marking length, width, and height.

85208-13_F20.EPS

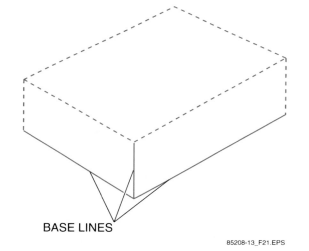

BASE LINES

85208-13_F21.EPS

Figure 21 Completing sketch.

Step 5 Add dimensions to the drawing as necessary. Make sure that all dimensions are clearly stated, and that no unnecessary information clutters the drawing.

Step 6 Add line numbers and direction of flow as necessary.

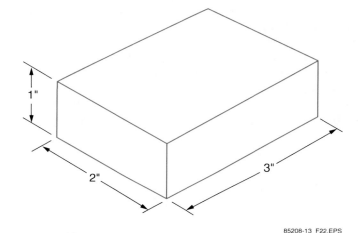

85208-13_F22.EPS

Figure 22 Sketching dimensions.

Step 7 Calculate needed pipe lengths from the orthographic drawing, then round up to the nearest foot or common metric dimension.

Step 8 Make a bill of materials to include needed pipe lengths, components such as valves, flanges, unions, vents, and drains, and other materials. All parts should have proper stock numbers.

Step 9 Make a title block with the following:
- Isometric number
- Line number(s)
- Reference number(s)
- A box marked "Drawn by" with your name.

Step 10 Carefully examine the finished sketch to make sure that any pipefitter could fabricate the line using the drawing as a guide.

Summary

A drawing is a detailed plan of a part or system. The drawing provides the information needed to build or install part of a system or the entire system. Pipefitters use drawings to install, troubleshoot, and repair piping.

The major parts of a drawing include title blocks, revision blocks, notes on the scale used, drawing indexes, and line lists. Drawings use several shortcuts, including symbols and abbreviations, with which the pipefitter must be familiar. Drawings contain notes and callouts that further inform the pipefitter as to how the assembly should be performed. Coordinates locate the pipes on an outline of the ship, referenced to several imaginary lines on the hull and superstructure.

Most drawings are made in the orthographic style. Orthographic drawings show the shape and size of a three-dimensional object through a series of related two-dimensional views. Each two-dimensional view shows only one side of the object. The views appear as though observer is looking at the object at a right angle to the surface of the object.

Commonly used drawings include a general arrangement plan, which is usually the top sheet of a set of drawings. Various detail sheets are based on the general arrangement plan. These include equipment arrangement drawings and piping and instrumentation drawings. Isometric drawings are used to show a three-dimensional view of piping layout. Spool drawings are used to make prefabricated parts that are then delivered to the site. Other drawings include pipe hanger and support drawings and as-built drawings. Equipment vendors supply some drawings.

The pipefitter must sometimes make freehand sketches to finish assembly of a pipe system. Sketches can be orthographic or isometric. Orthographic drawings are easier to make, but may require more than one view. Isometric drawings take more time and skill, but represent an object in a single view.

1. A part of another drawing, drawn at a larger scale than the original drawing, is called a _____.
   a. coordinate
   b. title block
   c. detail sheet
   d. note

2. The drawing index is developed from the _____.
   a. title block
   b. line list
   c. vendor list
   d. general arrangement plan

3. The drawing index is always on what page of the drawing set?
   a. First
   b. Second
   c. Last
   d. There is no specific page

4. A line list can be used to tell whether or not a line is insulated.
   a. True
   b. False

5. Which engineering drawing size is 17" × 22"?
   a. A
   b. B
   c. C
   d. D

6. The drawing scale is a ratio between the size of the drawing and the size of the _____.
   a. object being drawn
   b. vessel
   c. drawing paper
   d. vendor-supplied equipment

7. Dimension lines are bounded by _____.
   a. object lines
   b. cutting plane lines
   c. phantom lines
   d. extension lines

8. Interpretation of symbols used on a drawing is usually provided in the _____.
   a. revision block
   b. title block
   c. legend
   d. bill of materials

9. If an object such as a valve would show too much detail if drawn exactly, the person making the drawing usually _____.
   a. draws a circle with a note
   b. draws a symbol of the object
   c. adds a call-out at the bottom of the materials list
   d. adds a note at the top of the materials list

10. A revision block is used to record any changes made to the drawing.
    a. True
    b. False

11. A vertical line halfway between the bow and stern of a ship is called a _____.
    a. center line
    b. midship line
    c. base line
    d. buttock line

12. A line running athwartship is known as a _____.
    a. transverse frame line
    b. longitudinal frame line
    c. ship center line
    d. ship base line

13. In relation to the object being drawn, at what angle are orthographic projections made?
    a. 45 degrees
    b. 90 degrees
    c. 120 degrees
    d. 180 degrees

14. A plan view looks at an object from the _____.
    a. top
    b. side
    c. bottom
    d. front

15. A drawing resulting from an imaginary line cut through the ship is a(n) _____.

    a. plan view
    b. elevation view
    c. side view
    d. section view

16. On a piping and instrumentation drawing, valves are usually shown by _____.

    a. the letters VS
    b. notes
    c. symbols
    d. manufacturer stock numbers

17. How many pipelines are usually shown on a single isometric drawing?

    a. One
    b. Two
    c. Three
    d. Seven

18. Which of the following best describes the kind of piping shown on a spool drawing?

    a. Straight
    b. 90-degree angle
    c. Vendor supplied
    d. Prefabricated

19. An isometric sketch shows, at minimum, how many sides of the object being drawn?

    a. One
    b. Two
    c. Three
    d. Four

20. The left-side and the right-side isometric base lines should each form what kind of angle to the horizontal base line?

    a. 25-degree
    b. 30-degree
    c. 45-degree
    d. 90-degree

Trade Terms Introduced in This Module

Base line: A horizontal line at the bottom of the ship.

Bill of materials: A list of parts and materials shown on a drawing. Numbered entries on the bill of materials correspond to item (key) numbers located in the body of the drawing.

Buttock line: A line running parallel to and at specific distances from the center line of the ship.

Center line: A horizontal line running from bow to stern of the ship.

Elevation view: A drawing giving the height and length of an object, usually drawn looking toward the top of left side of the drawing.

Freeboard: The distance between the water line and the deck or weather deck of a ship.

Frame line: Lines perpendicular to the center line and base lines representing the locations of transverse frame members. Frames lines are used as measuring points for all distances fore and aft.

General arrangement plan: The master drawing that shows the layout of the complete unit to be constructed. It usually has the title block and a list of detailed drawings.

Graphic scale: A scale that shows the size relationship between the object represented in the drawing and the actual object.

Isometric drawing (ISO): A three-dimensional drawing of a piping system drawn to provide clarity.

Legend: An interpretation of symbols and abbreviations used on a drawing.

Line number: A group of abbreviations that specify size, service, material class/specification, insulation thickness, and tracing requirements of a given piping segment.

Load water line (LWL): The line at which a ship will float when loaded to its carrying capacity.

Longitudinal frame line: The line representing frame members running fore to aft along the length of the vessel.

Longitudinal frame: A shell, deck, or bulkhead stiffener running fore and aft.

Midship line: A vertical line halfway between the bow and stern. May be called the center line on some drawings.

Mold line (ML): The exact location at which a structure is installed.

Orthographic projection: The projection of a single view of an object.

Piping and instrumentation drawing (P&ID): A schematic flow diagram of a complete system or systems that shows function, instrument, valves, and equipment sequence.

Plan view: Drawing of an object or area from above it, looking down on the system.

Section view: A drawing of the view resulting from an imaginary line cut through the ship.

Sketch: A rough draft or free-hand drawing representing the primary features of an object.

Spool: A prefabricated segment of a pipe system or an ISO.

Transverse frame line: The line representing frame members running across the ship from port to starboard at specified intervals.

Transverse frames: Athwartships members forming the ship's ribs.

Water line: A line parallel to the ship base line and at various distances depending on the displacement of the ship.

Appendix A

PIPEFITTING ABBREVIATIONS

Adapter	ADPT
Air preheating	A
American Iron and Steel Institute	AISI
American National Standards Institute	ANSI
American Petroleum Institute	API
American Society for Testing and Materials International	ASTM
American Society of Mechanical Engineers	ASME
Ash removal water, sluice or jet	AW
Aspirating air	AA
Auxiliary steam	AS
Bench mark	BM
Beveled	B
Beveled end	BE
Beveled, both ends	BBE
Beveled, large end	BLE
Beveled, one end	BOE
Beveled, small end	BSE
Bill of materials	BOM
Blind flange	BF
Blowoff	BO
Bottom of pipe	BOP
Butt weld	BW
Carbon steel or cold spring	CS
Cast iron	CI
Ceiling	CLG
Chain operated	CO
Chemical feed	CF
Circulating water	CW
Cold reheat steam	CR
Compressed air	CA
Concentric or concrete	CONC
Condensate	C
Condenser air removal	AR
Continue, continuation	CONT
Coupling	CPLG

Detail	DET
Diameter	DIA
Dimension	DIM
Discharge	DISCH
Double extra-strong	XX STRG
Drain	DR
Drain funnel	DF
Drawing	DWG
Drip leg or dummy leg	DL
Ductile iron	DI
Dust collector	DC
Each	EA
Eccentric	ECC
Elbolet	EOL
Elbow	ELB
Electric resistance weld	ERW
Elevation	EL
Equipment	EQUIP
Evaporator vapor	EV
Exhaust steam	E
Expansion	EXP
Expansion joint	EXP JT
Extraction steam	ES
Fabrication (dimension)	FAB
Face of flange	FOF
Faced and drilled	F&D
Factory Mutual	FM
Far side	FS
Feed pump balancing line	FB
Feed pump discharge	FD
Feed pump recirculating	FR
Feed pump suction	FS
Female	F
Female	FM
Female pipe thread	FPT
Field support or forged steel	FS

85208-13_A01A.EPS

Field weld .... FW
Figure .... FIG
Fillet weld .... W
Finish floor .... F/F
Finish floor .... FIN FL
Finish grade .... FIN GR
Fitting .... FTG
Fitting makeup .... FMU
Fitting to fitting .... FTF
Flange .... FLG
Flat face .... FF
Flat on bottom .... FOB
Flat on top .... FOT
Floor drain .... FD
Foundation .... FDN
Fuel gas .... FG
Fuel oil .... FO

Gauge .... GA
Galvanized .... GALV
Gasket .... GSKT
Grating .... GRTG

Hanger .... HGR
Hanger rod .... HR
Hardware .... HDW
Header .... HDR
Heat traced, heat tracing .... HT
Heater drains .... HD
Heating system .... HS
Heating, ventilating, and air conditioning .... HVAC
Hexagon .... HEX
Hexagon head .... HEX HD
High point .... HPT
High pressure .... HP
Horizontal .... HORIZ
Hot reheat steam .... HR
Hydraulic .... HYDR

Increaser .... INCR
Input/output .... I/O
Inside diameter .... ID
Insulation .... INS
Invert elevation .... IE
Iron pipe size .... IPS
Isometric .... ISO
Issued for construction .... IFC

Lap weld .... LW
Large male .... LM
Length .... LG
Long radius .... LR
Long tangent .... LT
Long weld neck .... LWN
Low pressure .... LP
Low-pressure drains .... DR
Low-pressure steam .... LPS
Lubricating oil .... LO

Main steam .... MS
Main system blowouts .... BL
Makeup water .... MU
Male .... M
Malleable iron .... MI
Manufacturer .... MFR
Manufacturer's Standard Society .... MSS

National Pipe Thread .... NPT
Nipolet .... NOL
Nipple .... NIP
Nominal .... NOM
Not to scale .... NTS
Nozzle .... NOZ

Outside battery limits .... OSBL
Outside diameter .... OD
Outside screw and yoke .... OS&Y
Overflow .... OF

Pipe support .... PS
Pipe tap .... PT
Piping and instrumentation diagram .... P&ID
Plain .... P
Plain end .... PE
Plain, both ends .... PBE
Plain, one end .... POE
Point of intersection .... PI
Point of tangent .... PT
Pounds per square inch .... PSI
Process flow diagram .... PFD
Purchase order .... PO

Radius .... RAD
Raised face .... RF
Raised face slip-on .... RFSO

Raised face smooth finish .... RFSF
Raised face weld neck .... RFWN
Raw water .... RW
Reducer, reducing .... RED
Relief valve .... PRV-PSV
Ring-type joint .... RTJ
Rod hanger or right hand .... RH

Safety valve vents .... SV
Sanitary .... SAN
Saturated steam .... SS
Schedule .... SCH
Screwed .... SCRD
Seamless .... SMLS
Section .... SECT
Service and cooling water .... SW
Sheet .... SH
Short radius .... SR
Slip-on .... SO
Socket weld .... SW
Sockolet .... SOL
Stainless steel .... SS
Standard weight .... STD WT
Steel .... STL
Suction .... SUCT
Superheater drains .... SD
Swage .... SWG
Swaged nipple .... SN

Temperature or temporary .... TEMP
That is .... I.E.
Thick .... THK
Thousand .... M
Thread, threaded .... THRD
Threaded .... T
Threaded end .... TE
Threaded, both ends .... TBE
Threaded, large end .... TLE
Threaded, one end .... TOE
Threaded, small end .... TSE
Threadolet .... TOL
Top of pipe or top of platform .... TOP
Top of steel or top of support .... TOS
Treated water .... TW
Turbine .... TURB
Typical .... TYP

Underwriters Laboratories .... UL

Vacuum .... VAC
Vacuum cleaning VC
Vents V
Vitrified tile .... VT

Wall thickness or weight .... WT
Weld neck .... WN
Weldolet .... WOL
Well water .... WW
Wide flange .... WF

85208-13_A01C.EPS

Appendix B

PIPING SYMBOLS

TYPE OF FITTING		SOCKET	SCREWED	BUTT WELDED		FLANGED	
		SINGLE LINE	SINGLE LINE	DOUBLE LINE	SINGLE LINE	DOUBLE LINE	SINGLE LINE
90° ELL	TOP						
	SIDE						
	BOTTOM						
45° ELL	TOP						
	SIDE						
	BOTTOM						
TEE	TOP						
	SIDE						
	BOTTOM						
LATERAL	TOP						
	SIDE						
	BOTTOM						
REDUCERS AND SWAGES	CONC.						
	ECC.						

			SLIP-ON	WELD NECK	LAPPED	TONGUE & GROOVE	ORIFICE	BLIND
FLANGES	SINGLE LINE							
	DOUBLE LINE							
MISC.	SINGLE LINE							
	DOUBLE LINE							
			PIPE WELD	STUB IN	WELD SADDLE	PIPE CAP	UNION	CAP OR PLUG

85208-13_A02.EPS

Appendix C

VALVE AND SPECIAL SYMBOLS

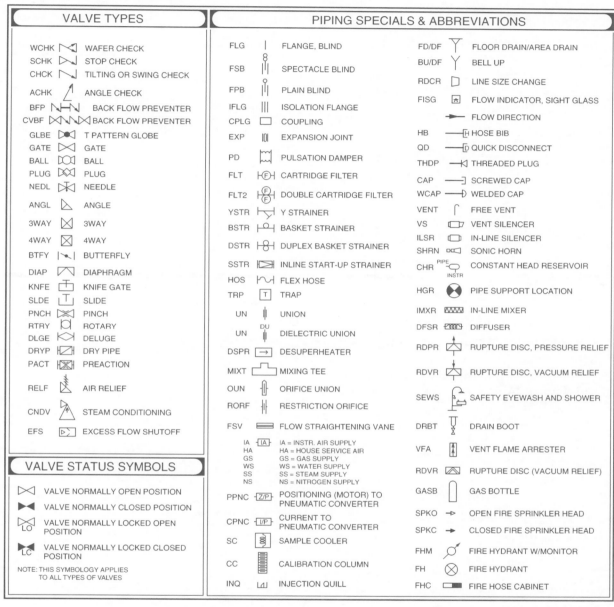

VALVE TYPES			PIPING SPECIALS & ABBREVIATIONS					
WCHK	WAFER CHECK		FLG	FLANGE, BLIND		FD/DF	FLOOR DRAIN/AREA DRAIN	
SCHK	STOP CHECK		FSB	SPECTACLE BLIND		BU/DF	BELL UP	
CHCK	TILTING OR SWING CHECK		FPB	PLAIN BLIND		RDCR	LINE SIZE CHANGE	
ACHK	ANGLE CHECK		IFLG	ISOLATION FLANGE		FISG	FLOW INDICATOR, SIGHT GLASS	
BFP	BACK FLOW PREVENTER		CPLG	COUPLING			FLOW DIRECTION	
CVBF	BACK FLOW PREVENTER		EXP	EXPANSION JOINT		HB	HOSE BIB	
GLBE	T PATTERN GLOBE		PD	PULSATION DAMPER		QD	QUICK DISCONNECT	
GATE	GATE		FLT	CARTRIDGE FILTER		THDP	THREADED PLUG	
BALL	BALL		FLT2	DOUBLE CARTRIDGE FILTER		CAP	SCREWED CAP	
PLUG	PLUG		YSTR	Y STRAINER		WCAP	WELDED CAP	
NEDL	NEEDLE		BSTR	BASKET STRAINER		VENT	FREE VENT	
ANGL	ANGLE		DSTR	DUPLEX BASKET STRAINER		VS	VENT SILENCER	
3WAY	3WAY		SSTR	INLINE START-UP STRAINER		ILSR	IN-LINE SILENCER	
4WAY	4WAY		HOS	FLEX HOSE		SHRN	SONIC HORN	
BTFY	BUTTERFLY		TRP	TRAP		CHR	CONSTANT HEAD RESERVOIR	
DIAP	DIAPHRAGM		UN	UNION		HGR	PIPE SUPPORT LOCATION	
KNFE	KNIFE GATE		UN	DIELECTRIC UNION		IMXR	IN-LINE MIXER	
SLDE	SLIDE		DSPR	DESUPERHEATER		DFSR	DIFFUSER	
PNCH	PINCH		MIXT	MIXING TEE		RDPR	RUPTURE DISC, PRESSURE RELIEF	
RTRY	ROTARY		OUN	ORIFICE UNION		RDVR	RUPTURE DISC, VACUUM RELIEF	
DLGE	DELUGE		RORF	RESTRICTION ORIFICE		SEWS	SAFETY EYEWASH AND SHOWER	
DRYP	DRY PIPE		FSV	FLOW STRAIGHTENING VANE				
PACT	PREACTION		IA	IA = INSTR. AIR SUPPLY		DRBT	DRAIN BOOT	
			HA	HA = HOUSE SERVICE AIR		VFA	VENT FLAME ARRESTER	
RELF	AIR RELIEF		GS	GS = GAS SUPPLY				
			WS	WS = WATER SUPPLY		RDVR	RUPTURE DISC (VACUUM RELIEF)	
CNDV	STEAM CONDITIONING		SS	SS = STEAM SUPPLY				
			NS	NS = NITROGEN SUPPLY		GASB	GAS BOTTLE	
EFS	EXCESS FLOW SHUTOFF		PPNC	POSITIONING (MOTOR) TO PNEUMATIC CONVERTER		SPKO	OPEN FIRE SPRINKLER HEAD	

VALVE STATUS SYMBOLS

	VALVE NORMALLY OPEN POSITION
	VALVE NORMALLY CLOSED POSITION
LO	VALVE NORMALLY LOCKED OPEN POSITION
LC	VALVE NORMALLY LOCKED CLOSED POSITION

NOTE: THIS SYMBOLOGY APPLIES TO ALL TYPES OF VALVES

CPNC	CURRENT TO PNEUMATIC CONVERTER	SPKC	CLOSED FIRE SPRINKLER HEAD
SC	SAMPLE COOLER	FHM	FIRE HYDRANT W/MONITOR
CC	CALIBRATION COLUMN	FH	FIRE HYDRANT
INQ	INJECTION QUILL	FHC	FIRE HOSE CABINET

85208-13_A03.EPS

Additional Resources

This module presents thorough resources for task training. The following resource material is suggested for further study.

Process Piping Drafting. Rip Weaver. Houston, TX: Gulf Publishing Company.

Figure Credits

Topaz Publications, Inc., Module opener

Courtesy of VT Halter Marine, Inc., Figures 1, 4, 5, 9–11, 15

Courtesy of Huntington Ingalls Shipbuilding, Figure 12–Figure 14

Utility Engineering Corp., a Zachry Group Company, Appendix C

NCCER CURRICULA — USER UPDATE

NCCER makes every effort to keep its textbooks up-to-date and free of technical errors. We appreciate your help in this process. If you find an error, a typographical mistake, or an inaccuracy in NCCER's curricula, please fill out this form (or a photocopy), or complete the online form at **www.nccer.org/olf**. Be sure to include the exact module ID number, page number, a detailed description, and your recommended correction. Your input will be brought to the attention of the Authoring Team. Thank you for your assistance.

Instructors – If you have an idea for improving this textbook, or have found that additional materials were necessary to teach this module effectively, please let us know so that we may present your suggestions to the Authoring Team.

NCCER Product Development and Revision
13614 Progress Blvd., Alachua, FL 32615

Email: curriculum@nccer.org
Online: www.nccer.org/olf

❏ Trainee Guide   ❏ Lesson Plans   ❏ Exam   ❏ PowerPoints   Other _____

Craft / Level: _____   Copyright Date: _____

Module ID Number / Title: _____

Section Number(s): _____

Description: _____

_____

_____

_____

_____

Recommended Correction: _____

_____

_____

_____

_____

Your Name: _____

Address: _____

_____

Email: _____   Phone: _____

Glossary

Acetone: A colorless organic solvent that is volatile and extremely flammable.

Acid: A chemical compound that reacts with and dissolves certain metals to form salts.

Actuator: The part of a regulating valve that converts electrical or fluid energy to mechanical energy to po-sition the valve.

Align: To make straight or to line up evenly.

Alloy: Any substance made up of two or more metals; two or more metals combined to make a new metal.

Angle valve: A type of globe valve in which the piping connections are at right angles.

ASTM International: Founded in 1898, a scientific and technical organization, formerly known as the American Society for Testing and Materials, formed for the development of standards on the characteristics and performance of materials, products, systems, and services.

Ball valve: A type of plug valve with a spherical disc.

Banded fitting: A pipe fitting that has a raised shoulder, or band, formed around the threaded opening to provide additional strength. Forged steel fittings are not banded.

Base line: A horizontal line at the bottom of the ship.

Bell-and-spigot: Pipe that has a bell, or en-largement, also called a hub, at one end of the pipe and a spigot, or smooth end, at the other end. The bell and spigot of two different pipes slide together to form a joint.

Bevel: An angle cut or ground on the end of a piece of solid material.

Bill of materials: A list of parts and materials shown on a drawing. Numbered entries on the bill of materi-als correspond item (key) numbers that in the body of the drawing.

Body: The main part of the valve. It contains the disc, seat, and valve ports. The body of the valve is directly connected to the piping by threaded, welded, or flanged ends.

Bonnet: The part of a valve containing the valve stem and packing.

Brazing: A method of joining metals with a nonferrous filler metal using heat above 842°F but be-low the melting point of the base metals being joined; also incorrectly known as hard soldering.

Brazing ring: Filler alloy shaped in a circle for insertion in a fitting. Brazing rings are used instead of conventional filler metals in certain situations.

Burn-through: A hole that is formed in a weld due to improper grinding or welding.

Butt weld: A method of joining in pipe in which the ends butt together without overlapping and are then welded using a filler metal.

Butterfly valve: A quarter-turn valve with a plate-like disc that stops flow when the outside area of the disc seals against the inside of the valve body.

Buttock line: A line running parallel to and at specific distances from the center line of the ship.

Capillary action: The tendency of a liquid to flow between narrow spaces. Capillary action is the gripping of a liquid to the walls of its container, combined with surface tension holding the sur-face of the liquid intact.

Cast iron: A brittle, hard alloy of carbon and iron.

Caustic: A material that is capable of burning or corroding by chemical action.

Cavitation: The result of pressure loss in liq-uid, producing bubbles (cavities) in vapor in liquid.

Cellular core wall: Plastic pipe wall that is low-density, lightweight plastic containing en-trained (trapped) air.

Center line: A horizontal line running from bow to stern of the ship.

Chamfer: To break the edge of construction material.

Check valve: A valve that allows flow in one direction only.

Chemically inert: Does not react with other chemicals.

Concentric reducer: A reducer that maintains the same center line between the two pipes that it joins.

Condensate: The liquid product of steam, caused by a loss in temperature or pressure.

Control valve: A globe valve automatically controlled to regulate flow through the valve.

Corrosive: Causing the gradual destruction of a substance by chemical action.

Cross: A fitting with four branches all at right angles to each other.

Cup depth: The distance that a tube inserts into a fitting, usually determined by a stop inside the fitting.

Cupronickel: An alloy consisting of copper and nickel with small amounts of manganese and iron.

Deformation: A change in the shape of a material or component due to an applied force or temperature.

Disc: Part of a valve used to control the flow of system fluid.

Ductile: A characteristic of metal that allows it to be fashioned into another form or drawn out.

Eccentric reducer: A reducer that displaces the center line of the smaller of the two joining pipes to one side.

Elastomeric: Elastic or rubberlike. Flexible, pliable.

Elevation view: A drawing giving the height and length of an object, usually drawn looking toward the top of left side of the drawing.

Fillet: A rounded internal corner or shoulder of filler metal. Often appears at the meeting point of a piece of tubing and a fitting when the joint is soldered or brazed.

Fillet weld: A weld with a triangular cross section joining two surfaces at right angles to each other.

Fit up: To put piping material in position to be welded together.

Flashback arrestor: A valve that prevents a flame from traveling back from the tip and into the hoses.

Flux: A chemical substance that prevents oxides from forming on the surface of metals as they are heated for soldering, brazing, or welding.

Forged steel: Steel that is formed by heating it to a high temperature, but without reaching the liquid state. While in a plastic-like state, the hot steel is then pressed, hammered, or rolled into the desired shape.

Frame line: Lines perpendicular to the center line and base lines representing the locations of transverse frame members. Frames lines are used as measuring points for all distances fore and aft.

Freeboard: The distance between the water line and the deck or weather deck of a ship.

Full-penetration weld: Complete joint penetration for a joint welded from one side only.

Galling: An uneven wear pattern between trim and seat that causes friction between the moving parts; deformity of the threads in which some of the thread material of one component is removed and transferred to the threads of the other, destroying their integrity.

Gasket: A device that is used to make a pressure-tight connection and that is usually in the form of a sheet or a ring.

Gate valve: A valve with a straight-through flow design that exhibits very little resistance to flow. It is normally used for open/shut applications.

General arrangement plan: The master drawing that shows the layout of the complete unit to be con-structed. It usually has the title block and a list of detailed drawings.

Globe valve: A valve in which flow is always parallel to the stem as it goes past the seat.

Graphic scale: A scale that shows the size relationship between the object represented in the drawing and the actual object.

Head loss: The loss of pressure due to friction and flow disturbances within a system.

Hydrostatic pressure test: The process in which a pipe is filled with water and all air is bled out from the highest and farthest points in the run.

Insert: A type of reducer that fits into the socket of a fitting to reduce the line size.

Interference fit: A fit that tightens as the pipe is pushed into the socket.

Isometric drawing (ISO): A three-dimensional drawing of a piping system drawn to provide clarity.

Kinetic energy: Energy of motion.

Lateral: A fitting or branch connection that has a side outlet that is any angle other than 90 degrees to the run.

Legend: An interpretation of symbols and abbreviations used on a drawing.

Length of effective thread: The portion of a threaded pipe that intrudes into a fitting or threaded component; also known as makeup or thread engagement.

Line number: A group of abbreviations that specify size, service, material class/specification, insulation thickness, and tracing requirements of a given piping segment.

Load water line (LWL): The line at which a ship will float when loaded to its carrying capacity.

Longitudinal frame: A shell, deck, or bulkhead stiffener running fore and aft.

Longitudinal frame line: The line representing frame members running fore to aft along the length of the vessel.

Makeup: The portion of a threaded pipe that intrudes into a fitting or threaded component; also known as the length of effective thread or thread engagement.

Malleable iron: Metallic fitting material that typically has a pressure rating of 125 to 150 psi when used for fittings. Due to their lack of strength and brittle nature, they are rarely used in maritime applications.

Material safety data sheet (MSDS): A document that describes the composition, characteristics, health hazards, and physical hazards of a specific chemical. It also contains specific information about how to safely handle and store the chemical and lists any special procedures or protective equipment required. The MSDS is being replaced by the safety data sheet (SDS) as part of the GHS program.

Midship line: A vertical line halfway between the bow and stern. May be called the center line on some drawings.

Mold line (ML): The exact location at which a structure is installed.

National Pipe Taper (NPT): The US standard for pipe threads. The other primary US standard, National Pipe Straight (NPS), has no ability to seal and is used in simple structural and assembly applications only.

Nonferrous: A group of metals and metal alloys that contain no iron.

Orthographic projection: The projection of a single view of an object.

Oxidation: The process by which the oxygen in the air combines with metal to produce tarnish and rust.

Oxide: A type of corrosion that is formed when oxygen combines with a base metal.

Packing: Material used to make a dynamic seal, preventing system fluid leakage around a valve stem.

Phonographic: When referring to the facing of a pipe flange, serrated grooves cut into the facing, resem-bling those on a phonograph record.

Piping and instrumentation drawing (P&ID): A schematic flow diagram of a complete system or systems that shows function, instrument, valves, and equipment sequence.

Plan view: Drawing of an object or area from above it, looking down on the system.

Plug: The moving part of a valve trim (plug and seat) that either opens or restricts the flow through a valve in accordance with its position relative to the valve seat, which is the stationary part of a valve trim.

Plug valve: A quarter-turn valve with a ported disc.

Positioner: A field-based device that takes a signal from a control system and ensures that the control de-vice is at the setting required by the control system.

Pounds per square inch gauge (psig): Amount of pressure in excess of the atmospheric pressure level.

Pressure differential: The difference in pressure between two points in a flow system. It is usually caused by frictional resistance to flow in the system.

Pressure rating: The maximum pressure at which a component or system may be operated continuously.

Pressure vessel: A metal container that can withstand high pressures.

Primer: A liquid applied to plastic pipe prior to solvent welding in order to pre-soften the pipe and ensure a strong solvent weld.

Reach rod: An extension device, usually flexible, that is attached to a control valve located in a hard-to-reach or hazardous location to enable remote manual control by an operator.

Relief valve: A valve that automatically opens when a preset amount of pressure is exerted on the valve disc.

Root opening: The space between the pipes at the beginning of a weld; the gap is usually 1/8 of an inch.

Seat: The part of a valve against which the disc presses to stop flow through the valve.

Section view: A drawing of the view resulting from an imaginary line cut through the ship.

Severe service: A high-pressure, high-temperature piping system.

Sketch: A rough draft or free-hand drawing representing the primary features of an object.

Socket weld: A method of joining pipe in which the pipe end is inserted into the recessed area of a fitting or valve and then welded using a filler metal.

Solid wall: Plastic pipe wall that does not contain trapped air.

Solvent weld: A joint created by joining two pipes using solvent cement that softens the material's surface.

Spool: A piping segment of a pipe system or an ISO.

Straight tee: A fitting that has one side outlet 90 degrees to the run.

Swage: A type of socketless fitting in which one side is larger than the other.

Tack-weld: A weld made to hold parts together in proper alignment until the final weld is made.

Takeout: Refers to the distance from the center of a fitting, such as an elbow, to the face or shoulder of the threaded opening.

Thermal transients: Short-lived temperature spikes.

Thermoplastic: Pipe that can be repeatedly softened by heating and hardened by cooling. When softened, thermoplastic pipe can be molded into desired shapes.

Thread angle: The angle formed between two inclined faces of a thread.

Thread crest: The top of a fabricated thread. The thread crest of an NPT thread, although it may appear sharp, actually has a very slight flattening at the crest.

Thread engagement: The portion of a threaded pipe that intrudes into a fitting or threaded component; also known as makeup or length of effective thread.

Throttling: The regulation of flow through a valve.

Torque: A twisting force used to apply a clamping force to a mechanical joint.

Transverse frame line: The line representing frame members running across the ship from port to star-board at specified intervals.

Transverse frames: Athwartships members forming the ship's ribs.

Trim: Functional parts of a pump or valve, such as seats, stem, and seals, that are inside the flow area.

Turbulence: The motion of fluids or gases in which velocities and pressures change irregularly.

Valve body: The part of a valve containing the passages for fluid flow, valve seat, and inlet and outlet con-nections.

Valve stem: The part of a valve that raises, lowers, or turns the valve disc.

Valve trim: The combination of the valve plug and the valve seat.

Water hammer: An increase in pressure in a pipeline caused by a sudden change in the flow rate. In a steam line, water hammer is caused by condensate blocking the flow of steam at a pipe bend.

Water line: A line parallel to the ship base line and at various distances depending on the displacement of the ship.

Wedge: The disc in a gate valve.

Wetting: A process that reduces the surface tension so that molten (liquid) filler flows evenly throughout the joint.

Wire drawing: The erosion of a valve seat under high velocity flow through which thin, wire-like gullies are eroded away.

Witness mark: A mark made for the purpose of determining the proper position of a pipe and fitting dur-ing the joining process. The witness mark is made in a position that will not be hidden when the pipe and fitting are assembled.

Working life: The time it takes for freshly mixed adhesive to begin to harden.

Yoke bushing: The bearing between the valve stem and the valve yoke.

Index

A

A Type metric copper pipe, (85201): 4
Abbreviation legends, (85208): 3
Abbreviations list
   pipefitting, (85208): 30–32
   piping specials, (85208): 35
Abrasive saws, (85202): 11
Acetone, (85204): 1, 25
Acetylene cylinders
   pressure-reducing regulator, (85204): 2
   ring guard caps, (85204): 12–14
   safety plugs, (85204): 18
   storage and transport, (85204): 1–2
   transducers, (85204): 18
   transportation and storage, (85204): 12
Acetylene cylinder valves, (85204): 2
Acid, (85201): 6, 15
ACR Type copper pipe
   applications, (85201): 3
   characteristics, (85201): 3–4
   color code, (85201): 5
   labeling, (85201): 5
   sizing, (85201): 4
Acrylic fiber gaskets, (85207): 43
Actuator, (85207): 1, 51. *See also* Valve actuators
Adhesives, fiberglass pipe, (85206): 2
Air-acetylene brazing equipment, (85204): 2, 4. *See also*
   Brazing equipment
Air-acetylene brazing torches
   lighting, (85204): 18–19
   torch tips, (85204): 19
Air motor-driven actuators, (85207): 27
Align, (85202): 1, 31
Aligning dogs, (85202): 17–18
Alignment tools, (85202): 20–21
Alloy, (85202): 6, 31; (85204): 8, 25
American Welding Society (AWS) brazing filler metals
   classification, (85204): 8
Amerilon® gaskets, (85207): 46
Angle globe valve, (85207): 11, 13–14, 51
Angle iron jigs, (85202): 16–17
Angle valve, (85207): 11, 51. *See also* Angle globe valve
Annealed (soft) copper pipe, (85201): 3
Approved by list on title block, (85208): 2
Asbestos gaskets, (85207): 45
ASTM International
   defined, (85202): 31; (85207): 51
   MSS marking system, (85207): 31
ASTM International standards
   A-197, (85202): 5
   B88 (imperial copper pipe), (85201): 4
   B88M (metric copper pipe), (85201): 4
   piping drawings, (85202): 5, 9
AWS brazing filler metals classification. *See* American
   Welding Society (AWS) brazing filler metals
   classification

B

Backing rings, selecting and installing, (85202): 14–16
Ball check valves, (85207): 22, 23
Ball valve, (85207): 6–8, 51
Banded fitting, (85205): 3, 19
Banded threaded pipe fittings, (85205): 3
Base line, (85208): 6, 29
BCuP-3, (85204): 8
Bell-and-spigot, (85206): 1, 17
Bell-and-spigot joints, Bondstrand® GRE pipe, (85206): 1,
   2–4
Bellows-type expansion joints, (85201): 11
Bell reducer, (85205): 6
Bevel, (85202): 1, 9, 31
Bevel-gear operators, (85207): 24, 25
Beveling
   for butt-welds, (85202): 5–6
   cutters for, (85202): 6
   grinders for, (85202): 6
   mechanical, (85202): 5–6
   oxyacetylene, (85202): 6, 9
   pipe bevellers for, (85202): 6, 9
   plastic pipe, (85206): 9
   thermal, (85202): 6, 9
Bill of materials
   butt weld piping drawings, (85202): 7
   common, (85208): 4–5
   defined, (85208): 29
   example, (85208): 7
   highlighted in text, (85208): 1
   socket weld piping drawings, (85203): 8
Black rubber gaskets, (85207): 44
Blind flange, (85207): 37
Body, (85207): 1, 51
Bondstrand® GRE pipe
   adhesive, (85206): 2
   advantages, (85206): 1, 2
   applications, (85206): 1
   assembly
      bell-and-spigot joints, (85206): 1, 2–4
      double-O-ring expansion coupling, (85206): 1, 2
      double-O-ring joints, (85206): 1, 2
      flanges, (85206): 1, 2
      Quick-Lock joints, (85206): 1, 2
      taper-taper joints, (85206): 1, 2
   fittings, (85206): 1, 2
Bonnet, (85207): 3, 51
Bow saws, (85202): 11
Branch connections
   butt weld fittings, (85202): 3
   threaded pipe fittings, (85205): 4–6
Brass, oxyacetylene brazing, (85204): 21
Brazing. *See also* Oxyacetylene brazing
   adhesion requirements, (85204): 1
   brass, (85204): 21
   clearance gap, (85204): 1
   defined, (85204): 1, 25

Brazing. *See also* Oxyacetylene brazing (*continued*)
    dissimilar metals, (85204): 21–22
    function, (85204): 1
    Iconel, (85204): 21
    materials used in
        brazing ring, (85204): 8, 9, 25
        filler metals, (85204): 8–9
        flux, (85204): 9, 25
    procedure
        horizontal pipe runs, (85204): 19–21
        pipe cutting and cleaning, (85204): 7–8
        pipe end preparations, (85204): 10–12
        vertical-up/-down joints, (85204): 21
    safety guidelines, (85204): 1–2, 16
    stainless steel, (85204): 21
    temperatures, (85204): 12
    valve bronze, (85204): 21
Brazing equipment
    air-acetylene setup, (85204): 2, 4
    oxyacetylene setup, (85204): 2–7
    welding goggles, (85204): 19
Brazing ring, (85204): 8, 9, 25
Break lines, (85208): 3
British Standard Pipe Parallel (BSPP), (85205): 1
British Standard Pipe Tapered (BSPT), (85205): 1
Bronze, oxyacetylene brazing, (85204): 21
BSPP. *See* British Standard Pipe Parallel (BSPP)
BSPT. *See* British Standard Pipe Tapered (BSPT)
BTK. *See* Buttock line (BTK)
B Type metric copper pipe, (85201): 4
Burn-through, (85202): 13, 31
Butterfly check valve, (85207): 22, 23
Butterfly valve, (85207): 14–16, 51
Buttock line (BTK), (85208): 6, 29
Butt weld fittings
    branch connections, (85202): 1–2, 3
    caps, (85202): 2, 3
    elbows, (85202): 1
    flanges
        characteristics, (85202): 2
        lap-joint, (85202): 2–3, 5
        reducing, (85202): 4
        weld-neck, (85202): 2, 4
    function, (85202): 1
    reducers, (85202): 2, 4
    return bends, (85202): 1
    symbols, (85202): 9
    takeouts, (85202): 11–13
Butt weld pipe fabrication
    alignment clamps and tools
        aligning dogs, (85202): 17–18
        angle iron jigs, (85202): 16–17
        cage clamps, (85202): 18–20
    alignment preparations
        beveling edges, (85202): 5–6, 9
        cleaning, (85202): 10
        cutting, (85202): 10–11
        grinding edges, (85202): 6–7, 9–11
    alignment procedures
        flange to pipe, (85202): 25–27
        45-degree elbow to pipe, (85202): 23–24
        90-degree angle, squaring, (85202): 25
        90-degree elbow to pipe, (85202): 24–25
        pipe-to-pipe, (85202): 21–23
        tees to pipe, (85202): 27
        valve to pipe, (85202): 27–28
    alignment requirements, (85202): 20
    alignment tools, (85202): 20–21

backing rings, selecting and installing, (85202): 14–16
    gap setting, (85202): 13
    pipe length calculations
        center-to-center method, (85202): 14
        center-to-face method, (85202): 14, 15
        face-to-face method, (85202): 14, 15
        length calculations, (85202): 13–14
        responsibility for, (85202): 13
        takeouts, (85202): 11–13
    root openings, (85202): 13
Butt weld piping drawings
    bill of materials, (85202): 7
    double-line drawings
        elevation view, (85202): 5
        function, (85202): 4
        plan view, (85202): 6
    isometric drawings, (85202): 4, 7
    piping symbols on, (85202): 4
        fitting symbols, (85202): 8
        valve symbols, (85202): 8
    single-line drawings
        elevation view, (85202): 5
        function, (85202): 4
        plan view, (85202): 6
    specifications, (85202): 5, 9
Butt welds
    applications, (85202): 1
    defined, (85201): 15
    illustrated, (85201): 2
    pipe wall thicknesses used with, (85202): 1
Butt weld valve symbols, (85202): 9

C

Cage clamps
    hydraulic, (85202): 18, 20
    lever-type, (85202): 18
    locking-screw, (85202): 18
    mechanical, (85202): 18
    straight pipe, (85202): 18–20
Capillary action, (85204): 1, 25
Caps
    butt weld fittings, (85202): 3
    PVC/CPVC pipe, (85206): 6
    socket weld, (85203): 2–3
    takeouts, (85202): 12
    threaded pipe fittings, (85205): 6
Carbon steel pipe, (85201): 2–4
Carburizing flame, (85204): 16, 18
Cast iron, (85205): 3, 19
Cast iron threaded pipe fittings, (85205): 3
Caustic, (85201): 6, 15
Cavitation, (85207): 36, 51
CC Type backing rings, (85202): 14–16
CCC Type backing rings, (85202): 14–16
Cellular core wall, (85206): 6, 17
Center line
    compartment identification in relation to, (85208): 7–8
    defined, (85208): 6, 29
    function, (85208): 3
Center-to-center method of calculating pipe length
    butt weld pipe fabrication, (85202): 14
    socket weld pipe fabrication, (85203): 5, 10
    threaded pipe fabrication, (85205): 9–12
Center-to-face method of calculating pipe length
    butt weld pipe fabrication, (85202): 14, 15
    socket weld pipe fabrication, (85203): 5–6, 11
    threaded pipe fabrication, (85205): 12–13
Ceramic gaskets, (85207): 45

Chain operators, valve actuators, (85207): 25
Chain-type channel-lock pliers, (85202): 16–17
Chamfer, (85206): 11, 17
Checked by list on title block, (85208): 2
Check valves, (85207): 9, 19–24, 34, 51
Chemically inert, (85206): 5, 17
Chemical piping systems, safety guidelines, (85201): 6
Chill rings. *See* Backing rings
Chlorinated polyvinyl chloride (CPVC) pipe. *See also* Plastic
    pipe
   advantages, (85206): 7–8
   applications, (85201): 2, (85206): 7
   fittings, (85206): 8
   joining, (85206): 8
Circular saws, (85202): 11
Coefficient of expansion by type of pipe, (85201): 10
Cold conservation insulation, (85201): 13
Coldsaws, (85202): 11
Cold spring gap, (85201): 12
Cold springing, (85201): 11–12
Color code
   copper pipe, (85201): 5
   fire suppression piping and fixtures, (85201): 1–2
   gaskets, (85207): 46
   hoses for brazing/welding, (85204): 3
   piping systems, (85201): 1
   PTFE tape, (85205): 13–14
   spiral-wound gaskets, (85207): 44
Compartment identification
   by compartment use, (85208): 8
   by deck number, (85208): 6
   by frame number, (85208): 7
   by position in relation to ship center line, (85208): 7–8
Compartment use designation, (85208): 8
Compressed air piping systems
   applications
      compressed air systems, (85201): 6–7
      socket weld systems, (85203): 1
      utility air systems, (85201): 6–7
   safety guidelines, (85201): 7
   socket weld fittings, (85203): 1
   types of
      instrument air, (85201): 6–7
      utility air, (85201): 7
Concentric reducer, (85203): 3, 21
Condensate, (85201): 8, 15
Cone valve, (85207): 7–8
Control valve, (85207): 1, 17, 20, 21, 51
Cooling water piping systems, (85201): 8
Copper pipe
   applications, (85201): 2
   categories by temper
      annealed (soft) copper, (85201): 3
      hard-drawn (hard) copper, (85201): 3
   color coding, by type, (85201): 5
   fabrication, (85201): 3
   fabrication standards, (85201): 4
   imperial, (85201): 4–5
   labeling, (85201): 5
   metric, (85201): 4–5
   types of, (85201): 3–5
   wall thicknesses by type, (85201): 3–5
Cork gaskets, (85207): 45
Corrosion in piping systems, (85201): 2
Corrosive, (85207): 1, 51
Corrugated metal gaskets, (85207): 43–44
   with asbestos inserted, (85207): 44
   with heat-resistant synthetic filler, (85207): 44

Couplings
   PVC/CPVC pipe, (85206): 6
   socket weld, (85203): 3, 4, 14
   threaded pipe, (85205): 6
CPVC pipe. *See* Chlorinated polyvinyl chloride (CPVC) pipe
Crosses
   butt weld, (85202): 2, 3
   defined, (85203): 21
   function, (85203): 1
   threaded pipe, (85205): 5
C Type backing rings, (85202): 14–16
C Type metric copper pipe, (85201): 4
CuNi. *See* Cupronickel (copper-nickel (CuNi))
Cup depth, (85204): 10, 25
Cupronickel (copper-nickel (CuNi)), (85201): 2, 15
Cupronickel (copper-nickel (CuNi)) pipe, (85201): 5–6
Customary system scale, (85208): 2
Cutters
   for beveling, (85202): 6
   pipe cutter, (85206): 9
   soft tubing cutters, (85206): 9
   tubing cutters, (85204): 7, (85206): 9
Cutting plane lines, (85208): 3

D

Deck number for compartment identification, (85208): 6
Deformation, (85207): 7, 51
Demineralized water piping systems, (85201): 8–9
Diamond port plug, (85207): 9
Diaphragm valves, (85207): 16–17, 18
Dielectric union, (85205): 7
Dimension lines, (85208): 3
Disc, (85207): 1, 51
Distilled water piping systems, (85201): 8
Double-O-ring expansion coupling, Bondstrand® GRE pipe,
    (85206): 1, 2
Double-O-ring joints, Bondstrand® GRE pipe, (85206): 1, 2
Drawings. *See also* Butt weld piping drawings; Socket weld
    piping drawings
   categories, major, (85208): 7
   compartment identification on
      by compartment use, (85208): 8
      by deck number, (85208): 6
      by frame number, (85208): 7
      by position in relation to ship center line, (85208): 7–8
   defined, (85208): 1
   function, (85208): 1, 6
   organization, (85208): 1
   orthographic projections
      common views illustrated, (85208): 9
      drawing, (85208): 8–9
      elevation view, (85208): 10, 12
      function, (85208): 8
      general arrangement (GA) plan, (85208): 10, 14–17
      plan views, (85208): 9, 11
      section views, (85208): 10, 13
   reference lines, common
      base line, (85208): 6
      buttock line (BTK), (85208): 6
      center line, (85208): 6
      freeboard, (85208): 6
      load water line (LWL), (85208): 6
      longitudinal frame line, (85208): 6
      midship line, (85208): 6
      mold line (ML), (85208): 6
      transverse frame line, (85208): 6
      water line, (85208): 6

Drawings (*continued*)
  sketches
    isometric sketches, (85208): 24–25
    orthographic sketches, (85208): 21, 24
  types of
    equipment arrangement, (85208): 17
    equipment vendor, (85208): 17, 21
    general arrangement (GA) plan, (85208): 10, 14–17
    isometric (ISO), (85208): 17, 19
    pipe support drawings and detail sheets, (85208): 21–23
    piping and instrumentation (P&ID), (85208): 17, 18
    spool, (85208): 17, 20
Drawings, common elements on
  bill of materials, (85208): 4–5, 7
  index, (85208): 1
  legends
    abbreviations, (85208): 3
    symbols, (85208): 3
  line list, (85208): 1
  lines
    break, (85208): 3
    center, (85208): 3
    commonly used, (85208): 3, 4
    cutting plane, (85208): 3
    dimension, (85208): 3
    hidden (invisible), (85208): 3
    leader, (85208): 3
    object (visible), (85208): 3
    phantom, (85208): 3
    reference lines, (85208): 5–6
  measurements, (85208): 2–3
  notes, (85208): 4, 5
  revision block, (85208): 4, 6
  scale
    customary system, (85208): 2
    graphic, (85208): 3
    metric system, (85208): 2
    N.T.S. (not to scale), (85208): 3
    on title block, (85208): 1
    U.N. (unless noted otherwise), (85208): 2
  symbols
    fittings, (85208): 33–34
    legend, (85208): 3
    pipe hangers and supports, (85208): 22
    specials, (85208): 35
    valves, (85208): 35
  title block, information on
    approved by, (85208): 2
    checked by, (85208): 2
    drawing title, (85208): 1
    drawn by, (85208): 2
    key plan, (85208): 2
    project number, (85208): 1
    revision number, (85208): 1–2
    scale, (85208): 1
    sheet number, (85208): 1
    sheet size, (85208): 1, 2
    zone, (85208): 2
Drawn by list on title block, (85208): 2
Ductile, (85205): 3, 19
DWV Type copper pipe
  applications, (85201): 3
  color code, (85201): 5
  labeling, (85201): 5
  wall thicknesses, (85201): 3

E

Eccentric reducer, (85203): 3, 21
Elastomeric, (85207): 5, 51
Elbolet
  butt weld, (85202): 3
  socket weld, (85203): 3, 4
Elbows
  angles and radius, (85202): 1, 2
  butt weld, (85202): 1
  function, (85202): 1
  PVC/CPVC pipe, (85206): 6
  threaded pipe, (85205): 4–5
Electric actuators, (85207): 27
Elevation view
  butt weld piping drawings, (85202): 5
  defined, (85208): 29
  orthographic projections, (85208): 10, 12
  reference lines, (85208): 9
  socket weld piping drawings, (85203): 7
End-entry ball valves, (85207): 7
EPDM gaskets, (85207): 45
Equal percentage plug, (85207): 13
Equipment arrangement drawings, (85208): 17
Equipment vendor drawings, (85208): 17, 21
Expansion joints, (85201): 11
Expansion loops, (85201): 10–11
Expansion U-loops, (85201): 11
Extra-strong wall galvanized pipe, (85201): 2

F

Face-to-face method of calculating pipe length
  butt weld pipe fabrication, (85202): 14, 15
  socket weld pipe fabrication, (85203): 6, 11
  threaded pipe fabrication, (85205): 13
Feather edge, (85202): 5, 9
Fiberglass gaskets, (85207): 43
Fiberglass insulation, (85201): 13
Fiberglass pipe. *See also* Bondstrand® GRE pipe
  adhesives, (85206): 2
  applications, (85201): 2, (85206): 1
  benefits, (85206): 1
  fabrication, (85206): 1
  sizes, (85206): 1
Fiberglass pipe shaver, (85206): 4
Filler metals, brazing, (85204): 8–9
Fillet, (85204): 20, 25
Fillet weld, (85203): 1, 21
Fire resistance, plastic pipe, (85201): 2
Fire suppression piping and fixtures color codes, (85201): 1–2
Fittings. *See also* Butt weld fittings; Socket weld fittings; Threaded pipe fittings
  banded fitting, (85205): 3, 19
  banded threaded pipe fittings, (85205): 3
  Bondstrand® GRE pipe, (85206): 1, 2
  grooved fittings, (85201): 2
  PVC/CPVC pipe, (85206): 5–6, 7
  symbols, (85202): 8, (85208): 33–34
Fit up, (85202): 1, 31
Flames, oxyacetylene, (85204): 16, 18
Flammable liquid piping systems, (85201): 6–7
Flanged joint, (85207): 1
Flange facing finishes, (85207): 38, 39, 40
Flange facings
  flat-face flanges, (85207): 39
  male and female flanges, (85207): 41
  raised-face flanges, (85207): 38

ring joint type (RJT) flanges, (85207): 39, 41
tongue and groove flanges, (85207): 41
Flange gaskets, types of
flat ring gaskets, (85207): 46–47
full-face gaskets, (85207): 46–47
Flanges
Bondstrand® GRE pipe joints, (85206): 1, 2
butt weld
basics, (85202): 2
lap-joint, (85202): 2–3, 5
reducing, (85202): 4
weld-neck, (85202): 2, 4
function, (85207): 35
threaded pipe fittings, (85205): 8
types of
blind flange, (85207): 37
lap-joint flange, (85207): 37–38
silver-brazed flange, (85207): 38
slip-on flange, (85207): 35, 36
slip-on reducing, (85207): 35–36
socket weld, (85203): 3, 6
socket weld flange, (85207): 37
threaded flange, (85207): 37
Van Stone flanges, (85202): 2–3
weld-neck flanges, (85202): 2, 4, (85207): 35
Flange-to-pipe alignment
butt weld pipe fabrication, (85202): 25–27
socket weld pipe fabrication, (85203): 11–13
Flange-to-vertical pipe alignment, socket weld pipe
fabrication, (85203): 13–14
Flashback arrestor, (85204): 4, 25
Flashbacks, (85204): 5
Flat-face flanges, (85207): 39
Flat metal gaskets with asbestos material, (85207): 44
Flat ring gaskets, (85207): 46–47
Flux
brazing
function, (85204): 9
stainless steel, (85204): 21
defined, (85204): 25
function, (85204): 1
soldering, (85204): 9
Foot valves, (85207): 23–24
Forged steel, (85205): 3, 19
Forged steel threaded pipe fittings, (85205): 4
45-degree elbows, (85202): 1, 11–12, (85203): 1
45-degree elbow-to-pipe alignment
butt weld pipe fabrication, (85202): 23–24
socket weld pipe fabrication, (85203): 14
Frame lines, (85208): 6, 9, 29
Frame number for compartment identification, (85208): 7
Framing squares, aligning pipe with, (85202): 20–25, 27–28
Freeboard, (85208): 6, 29
Fuel gas cylinders, storing, (85204): 2
Fuel gas hoses, (85204): 3
Fuel oil piping systems, (85201): 6–7
Full-face gaskets, (85207): 46–47
Full-penetration weld, (85202): 13, 31
Full-port ball valves, (85207): 7

G

Galling, (85205): 19, (85207): 9, 51
Galvanized (carbon steel) pipe, (85201): 2–3
Gap-A-Let®, (85203): 5, 8
GA plan. See General arrangement (GA) plan
Gaskets
color code, (85207): 46
defined, (85207): 1, 51

materials used in, (85207): 41–42
replacements for, (85207): 41
types of
acrylic fiber gaskets, (85207): 43
Amerilon® gaskets, (85207): 46
asbestos gaskets, (85207): 45
ceramic gaskets, (85207): 45
cork gaskets, (85207): 45
EPDM gaskets, (85207): 45
fiberglass gaskets, (85207): 43
flange gaskets, (85207): 46–47
flat ring gaskets, (85207): 46–47
full-face gaskets, (85207): 46–47
graphite-impregnated gaskets, (85207): 46
Gylon® gaskets, (85207): 46
metal gaskets, (85207): 43–44
natural rubber gaskets, (85207): 44
neoprene gaskets, (85207): 45
nitrile gaskets, (85207): 45
PTFE gaskets (Teflon®), (85207): 42–43
silicone gaskets, (85207): 45
soft metal gaskets, (85207): 46
vinyl gaskets, (85207): 45
Viton® gaskets, (85207): 45–46
Gate valve fluid-control elements
parallel-disc, (85207): 4–5
wedges
flexible, (85207): 2, 3
solid, (85207): 2, 3
split, (85207): 2, 3–4
Gate valves, (85207): 1–5, 29–30, 51
Gear operators, actuators
bevel-gear operators, (85207): 24, 25
spur-gear operators, (85207): 24, 25
worm-gear operators, (85207): 24–25, 25
General arrangement (GA) plan, (85208): 1, 14–17
General arrangement plan, (85208): 29
General notes, (85208): 4, 5
GHS. See Globally Harmonized System of Classification
(GHS)
Glass reinforced epoxy (GRE), (85206): 1
Globally Harmonized System of Classification (GHS),
(85201): 6
Globe valves
advantages/disadvantages, (85207): 11
applications, (85207): 1
defined, (85207): 51
disc and seat arrangements, (85207): 13–14
installation, (85207): 14, 30
markings, (85207): 31
Graphic scale, (85208): 3, 29
Graphite-impregnated gaskets, (85207): 46
GRE. See Glass reinforced epoxy (GRE)
Grinders
for beveling, (85202): 6
operation guidelines, (85202): 9–11
safety guidelines, (85202): 7, 9
Guillotine saws, (85202): 11
Gylon® gaskets, (85207): 46

H

Hacksaws, (85204): 7
Hard (hard-drawn) copper pipe, (85201): 3
Head loss, (85207): 11, 51
Heat conservation insulation, (85201): 13
Hidden (invisible) lines, (85208): 3
High temperature/high pressure gaskets, (85207): 41

High temperature/low pressure gaskets, (85207): 41
High-test, reinforced rubber gaskets, (85207): 44
Hi-Lo gauge, (85202): 20–21, 22
Hydraulic actuators, (85207): 27
Hydraulic cage clamps, (85202): 18, 20
Hydrostatic pressure test, (85206): 14, 17

I

Iconel, oxyacetylene brazing, (85204): 21
Imperial copper pipe, (85201): 4–5
Insert, (85203): 3, 4, 21
Instrument air piping systems, (85201): 6–7
Interference fit, (85206): 12, 17
Invisible (hidden) lines, (85208): 3
ISO drawings. *See* Isometric (ISO) drawings
Isometric (ISO) drawings
    butt weld piping drawings, (85202): 4, 7
    defined, (85208): 29
    example, (85208): 19
    function, (85208): 17
    highlighted in text, (85208): 10
    markings, (85208): 17
    socket weld piping drawings, (85203): 4, 8
    symbols, (85208): 17
Isometric sketches, (85208): 24–25

J

Joint compound, threaded pipe fabrication, (85205): 13,
    14–15
Joint tape, (85205): 13–14

K

Key plan on title block, (85208): 2
Kinetic energy, (85207): 11, 51
Knife gate valve, (85207): 5–6
K Type copper pipe
    applications, (85201): 3
    color code, (85201): 5
    labeling, (85201): 5
    wall thicknesses, (85201): 3

L

Labeling. *See* Markings
Laminated metal gaskets, (85207): 44
Lands, (85202): 5, 9
Lap-joint flanges, (85202): 2–3, 5, (85207): 37–38
Lateral branch connections, (85202): 2, 3
Laterals, (85202): 1, 31
Latrolet, (85203): 3, 4
Leader lines, (85208): 3
Legends
    abbreviations, (85208): 3
    defined, (85208): 29
    highlighted in text, (85208): 1
    symbols, (85208): 3
Length of effective thread, (85205): 9, 19
Levels, aligning pipe using, (85202): 20, 25–26
Lever-type cage clamps, (85202): 18
Lift check valve, (85207): 21–22, 23
Lift-type non-lubricated valves, (85207): 10
Line list information on drawings, (85208): 1
Line number, (85208): 1, 29
Lines on drawings, common
    break, (85208): 3
    center, (85208): 3
    cutting plane, (85208): 3
    dimension, (85208): 3

    hidden (invisible), (85208): 3
    leader, (85208): 3
    object (visible), (85208): 3
    phantom, (85208): 3
Load water line (LWL), (85208): 6, 29
Locking-screw cage clamps, (85202): 18
Longitudinal frame, (85208): 6, 29
Longitudinal frame line, (85208): 6, 29
Long-radius 90-degree elbows
    basics, (85202): 1–2
    takeouts, (85202): 11–13
Low temperature/high pressure gaskets, (85207): 42
Low temperature/low pressure gaskets, (85207): 42
L Type copper pipe
    applications, (85201): 3
    color code, (85201): 5
    labeling, (85201): 5
    wall thicknesses, (85201): 3
Lubricated plug valves, (85207): 9–10

M

Makeups, (85205): 11, 19
Male and female flanges, (85207): 41
Malleable iron, (85205): 3, 19
Malleable iron threaded pipe fittings, (85205): 3–4
Manufacturers Standardization Society (MSS), (85207): 31
Markings
    color code
        copper pipe, (85201): 5
        fire suppression piping and fixtures, (85201): 1–2
        gaskets, (85207): 46
        hoses for brazing/welding, (85204): 3
        piping systems, (85201): 1
        PTFE tape, (85205): 13–14
        spiral-wound gaskets, (85207): 44
    isometric (ISO) drawings, (85208): 17
    threaded pipe fabrication, (85205): 9
    valves
        bridgewall marking, (85207): 29
        common, (85207): 32
        flow direction arrow, (85207): 15, 29
        globe valve, (85207): 31
        installation and, (85207): 29
        rating designation, (85207): 32
        schematic symbols, (85207): 32–33
        size designation, (85207): 33
        standards, (85207): 31
        thread markings, (85207): 33
        trim identification, (85207): 32
Material Safety Data Sheet (MSDS), (85201): 6, 15
Mechanical beveling, (85202): 5–6
Mechanical cage clamps, (85202): 18
Medical Gas Type K copper pipe, (85201): 4
Medical Gas Type L copper pipe, (85201): 4
Metal gaskets, (85207): 43–44
Metric copper pipe, (85201): 4–5
Metric system scale, (85208): 2
Midship line, (85208): 6, 29
Mild steel pipe, joining, (85201): 2
Mineral wool insulation, (85201): 13
ML. *See* Mold line (ML)
Mold line (ML), (85208): 6, 29
Monel™ pipe, (85201): 6
Motor-actuated valves (MOVs), (85207): 27
MOVs. *See* Motor-actuated valves (MOVs)
MSDS. *See* Material Safety Data Sheet (MSDS)
MSS. *See* Manufacturers Standardization Society (MSS)

M Type copper pipe
  applications, (85201): 3
  color code, (85201): 5
  labeling, (85201): 5
  wall thicknesses, (85201): 3
Multi-flame torch tip, (85204): 6, 20
Multiport valves, (85207): 9

N

National Pipe Straight (NPS), (85205): 1
National Pipe Thread (NPT), (85205): 1–2, 19
Natural rubber gaskets, (85207): 44
Navy, US, (85201): 2
Needle plug, (85207): 14
Needle valves, (85207): 17, 19
Neoprene gaskets, (85207): 45
Neutral flame, (85204): 16, 18
Nickel-copper (NiCu) pipe, (85201): 6
Nickel-copper, (NiCu), pipe, nickel–, pipe
90-degree elbows
  function, (85203): 1
  squaring
    butt weld pipe fabrication, (85202): 25
    socket weld pipe fabrication, (85203): 10
  types of, (85202): 1
90-degree elbow-to-pipe alignment
  butt weld pipe fabrication, (85202): 24–25
  socket weld pipe fabrication, (85203): 10–11
Nipples, threaded pipe fittings, (85205): 7–8, 9
Nitrile gaskets, (85207): 45
Nominal size designation, (85201): 2–3
Nonferrous, (85204): 1, 25
Non-lubricated plug valves, (85207): 10
Nonrising stem, (85207): 5, 6
Notes on drawings
  general, (85208): 4, 5
  location, (85208): 4
  specifications, (85208): 4
Not to scale (N.T.S.), (85208): 3
NOV Fiber Glass Systems, (85206): 1
NPS. *See* National Pipe Straight (NPS)
NPT. *See* National Pipe Thread (NPT)
N.T.S. *See* Not to scale (N.T.S.)

O

Object (visible) lines, (85208): 3
Offsets, threaded pipe fittings, (85205): 4–5
180-degree returns
  basics, (85202): 1, 2
  takeouts, (85202): 12
O-rings, (85203): 1
Orthographic projections
  defined, (85208): 29
  elevation view, (85208): 10, 12
  function, (85208): 8
  plan views, (85208): 9, 11
  section views, (85208): 10, 13
Orthographic sketches, (85208): 21, 24
OS&Y stem. *See* Outside screw-and-yoke (OS-Y) stem
Outside screw-and-yoke (OS&Y) stem, (85207): 5, 6
Oxidation, (85204): 9, 25
Oxides, (85202): 6, 31
Oxidizing flame, (85204): 16, 18
Oxyacetylene beveling, (85202): 6, 9
Oxyacetylene brazing. *See also* Brazing
  equipment set-up, (85204): 12–15
  of horizontal pipe runs, (85204): 19–21
  lighting the torch, (85204): 15–18

safety guidelines, (85204): 16
  of vertical-up-/-down joints, (85204): 21
Oxyacetylene brazing equipment
  cup-type striker, (85204): 6–7
  cylinders
    clamshell caps, (85204): 12–14
    empty, (85204): 12
    hoses, (85204): 3
    labeling, (85204): 12
    regulators, (85204): 2–3, 5
    ring guard caps, (85204): 12–14
    safety plugs, (85204): 18, 19
    transducers, (85204): 18, 19
    transportation and storage, (85204): 1–2, 12
  flashback arrestors, (85204): 5
  portable, (85204): 4
  safety guidelines, (85204): 1–2
  torch handles and tips, (85204): 4–6, 16
  torch tip cleaners, (85204): 6, 7
  torch wrench, (85204): 3–4, 6
  typical setup, (85204): 2, 3
Oxyacetylene torches
  flame types, (85204): 16, 18
  handles, (85204): 4–6
  lighting, (85204): 15–17
Oxyacetylene torch tips
  multi-flame/rosebud, (85204): 6, 20
  operational data, (85204): 16
  selecting, (85204): 4–6
  tip cleaners, (85204): 6, 7
Oxygen cylinders
  storage and handling, (85204): 2
  transducers, (85204): 19
  transportation and storage, (85204): 12
Oxygen cylinder valves, (85204): 2
Oxygen hoses, (85204): 3

P

P&ID drawings. *See* Piping and instrumentation (P-ID) drawings
Packing, (85207): 5, 51
Parallel-disc gate valves, (85207): 4–5
Personal protection insulation, (85201): 12
Phantom lines, (85208): 3
Phonographic, (85207): 30, 51
Pipe. *See also* Fiberglass pipe; Plastic pipe
  cleaning
    for brazing, (85204): 7–8
    for butt welds, (85202): 6–7, 9–11
  cutting, (85202): 10–11, (85204): 7
  joining methods, by type of pipe, (85201): 2
  physical movement stress, (85201): 9
  standard weight, (85201): 2
  thermal expansion stress
    calculating, (85201): 9–10
    causes, (85201): 9
    compensating for, (85201): 10–12
    results, (85201): 9
Pipe bevellers, (85202): 6, 9
Pipe cutters, (85206): 9
Pipe dope. *See* Joint compound
Pipe-end-prep lathes, (85202): 6
Pipe insulation
  function, (85201): 12
  materials used in
    fiberglass, (85201): 13
    mineral wool, (85201): 13

Pipe insulation (*continued*)
  types of
    cold conservation, (85201): 13
    heat conservation, (85201): 13
    personal protection, (85201): 12
Pipe support drawings and detail sheets, (85208): 21
Pipe thread engagement, (85205): 2–3
Pipe threads
  defined, (85205): 1
  standards
    British Standard Pipe Parallel (BSPP), (85205): 1
    British Standard Pipe Tapered (BSPT), (85205): 1
    National Pipe Straight (NPS), (85205): 1
    National Pipe Thread (NPT), (85205): 1–2
  tapered, (85205): 2–3
  thread angle, (85205): 1, 19
  thread crest, (85205): 1, 19
Pipe-to-pipe alignment
  butt weld pipe fabrication, (85202): 21–23
  socket weld pipe fabrication, (85203): 14–15
Pipe wraparound, (85206): 8, 9
Pipe wrench, (85205): 15
Piping and instrumentation (P&ID) drawings, (85208): 10,
    17, 18, 29
Piping systems
  color coding, (85201): 1
  corrosion in, (85201): 2
  defined, (85201): 1
  noise, compensating for, (85201): 11
  safety guidelines when working with
    chemical systems, (85201): 6
    compressed air systems, (85201): 7
  schematics
    compressed air system, (85201): 7
    fuel oil piping system, (85201): 7
    potable water piping system, (85201): 9
    steam piping system, (85201): 9
  types and functions of , (85201): 1
  vibration, compensating for, (85201): 11
Plan view
  butt weld piping drawings, (85202): 6
  defined, (85208): 29
  general arrangement (GA) plan, (85208): 10, 14–17
  key plan shown, (85208): 2
  orthographic projections, (85208): 9, 11
  socket weld piping drawings, (85203): 7
Plastic pipe
  advantages/disadvantages, (85206): 5
  applications, (85201): 2
  characteristics, (85206): 5
  fire resistance, (85201): 2
  fittings, (85206): 5
  sizes, (85206): 5
  storage, (85206): 5
Plastic pipe assembly
  pre-fabrication
    beveling, (85206): 9
    cleaning, (85206): 9, 10
    cutting, (85206): 8, 9
    deburring, (85206): 9, 10
    marking, (85206): 8, 9
    priming, (85206): 9, 10
    reaming, (85206): 8
  pressure testing post-, (85206): 15
  solvent cementing
    cure times, (85206): 12
    process, (85206): 10–13
    products, (85206): 8

set-up times, (85206): 11
support spacing, (85206): 12, 14
Plugs
  defined, (85207): 51
  function, (85207): 1
  globe valve, (85207): 14
  threaded pipe fittings, (85205): 6
Plug-type disc, (85207): 13
Plug valve, (85207): 8–10, 51
Pneumatic actuators, (85207): 27
Polyvinyl chloride (PVC) pipe. *See also* Plastic pipe
  applications, (85201): 2, (85206): 6
  characteristics, (85206): 6
  fittings, (85206): 6
  joining, (85206): 7
Positioner, (85207): 51
Potable water, (85201): 1, 8
Potable water piping system, (85201): 8
Pounds per square inch gauge (psig), (85201): 7, 15
Pressure differential, (85207): 36, 51
Pressure ratings
  defined, (85206): 17
  labeling, (85206): 5
  socket weld fittings, (85203): 1
  working pressure rating
    annealed (soft) copper pipe, (85201): 3
    hard (hard-drawn) copper pipe, (85201): 3
Pressure-relief valves, (85207): 19, 21, 22, 34, 51
Pressure vessel, (85207): 37, 51
Primer, (85206): 9, 17
Project number on title block, (85208): 1
psig. *See* Pounds per square inch gauge (psig)
PTFE gaskets (Teflon®), (85207): 42–43
PTFE tape, (85205): 13–14
PVC/CPVC pipe. *See also* Plastic pipe
  applications, (85201): 2, (85206): 6–7
  dimensions, (85206): 5–6
  fittings, (85206): 5–6, 7, 8
  joining, (85206): 7, 8
  labeling, (85206): 5, 6
  schedules, (85206): 5–6
  solvent cementing products, (85206): 8
PVC pipe. *See* Polyvinyl chloride (PVC) pipe

Q

Quick-Lock joints, Bondstrand® GRE pipe, (85206): 1, 2

R

Ratchet shears, (85206): 9
Rating designation markings, valves, (85207): 32
Reach rod, (85207): 24, 27–28, 51
Reamers, (85204): 7–8, (85206): 8–9
Rectangular plug, (85207): 9
Red rubber gaskets, (85207): 44
Reducers
  butt weld, (85202): 2, 4
  PVC/CPVC pipe, (85206): 6
  socket weld, (85203): 3
Reducing bushings, (85205): 7
Reducing coupling, (85205): 6
Reducing elbows, (85202): 1
Reducing flanges, (85202): 4
Reference lines, common
  base line, (85208): 6, 29
  buttock line (BTK), (85208): 6, 29
  center line, (85208): 6, 29
  freeboard, (85208): 6, 29
  load water line (LWL), (85208): 6, 29

longitudinal frame line, (85208): 6, 29
midship line, (85208): 6, 29
mold line (ML), (85208): 6, 29
in relation to the hull, (85208): 9
transverse frame line, (85208): 6, 29
water line, (85208): 6
Regulating valves
directional
ball check valves, (85207): 22, 23
butterfly check valve, (85207): 22, 23
check valve, (85207): 19, 51
foot valves, (85207): 23–24
lift check valve, (85207): 21–22, 23
swing check valve, (85207): 20–21, 23
stop/start
angle globe valve, (85207): 11, 13–14, 51
butterfly valve, (85207): 14–16, 51
diaphragm valves, (85207): 16–17, 18
globe valve, (85207): 11, 13–14, 51
needle valves, (85207): 17, 19
Y-type, (85207): 14
symbols, (85207): 34
Reinforced rubber gaskets, (85207): 44
Relief valve, (85207): 19, 34, 51
Return bends
butt weld fittings, (85202): 1
threaded pipe fittings, (85205): 4–5
Revision number on title block, (85208): 1–2
Rigging valves, (85207): 28–29
Ring joint metal gaskets, (85207): 43, 44
Ring joint type (RJT) flanges, (85207): 39, 41
Rising stem, (85207): 5, 6
RJT flanges. See Ring joint type (RJT) flanges
Root face, (85202): 5
Root opening, (85202): 11, 13, 31
Rosebud torch tip, (85204): 6
Round port plug, (85207): 9
Rubber gaskets, (85207): 44

S

Saddles, branch connections, (85202): 2, 3
Safety
brazing, (85204): 1–2
grinders, (85202): 7
grinders use, (85202): 9
Material Safety Data Sheet (MSDS), (85201): 6, 15
oxyacetylene brazing, (85204): 16
oxyacetylene equipment, (85204): 1–2
valve storage and handling, (85207): 28
welding, (85204): 19
Safety data sheet (SDS), (85201): 6
Safety valves, (85207): 19, 22
Scale
customary system, (85208): 2
graphic, (85208): 3
metric system, (85208): 2
N.T.S. (not to scale), (85208): 3
on title block, (85208): 1
U.N. (unless noted otherwise), (85208): 2
SDS. See Safety data sheet (SDS)
Seat, (85207): 1, 51
Section view
defined, (85208): 29
frame lines, (85208): 9
orthographic projections, (85208): 10, 13
Serrated metal gaskets, (85207): 43
Severe service, (85207): 35, 51
Sheet number on title block, (85208): 1

Sheet size on title block, (85208): 1, 2
Short-radius 90-degree elbows
basics, (85202): 1–2
takeouts, (85202): 11–13
Silicone gaskets, (85207): 45
Silver-brazed flange, (85207): 38
Size designation markings, valves, (85207): 32
Slip-on flange, (85207): 35, 36
Slip-on reducing flange, (85207): 35–36
Slip-on sleeve joints, (85203): 3, 4, 16–17
Socket weld fittings
caps, (85203): 2–3
center-to-end dimensions, (85203): 8
classes, (85203): 1
common types, (85203): 1
compressed air systems, (85203): 1
couplings, (85203): 3, 4, 14
dimensions, (85203): 1, 2
elbolet, (85203): 3, 4
flanges, (85203): 3, 6, (85207): 37
latrolet, (85203): 3, 4
laying length, (85203): 8
pressure ratings, (85203): 1
reducers, (85203): 3, 4
slip-on sleeve joints, (85203): 3, 4, 16–17
sockolet, (85203): 3, 4
takeouts, (85203): 5, 8
unions, (85203): 1–2
Socket weld pipe fabrication
alignment preparations, (85203): 7–10
alignment procedures
flange-to-pipe alignment, (85203): 11–13
flange-to-vertical pipe alignment, (85203): 13–14
45-degree elbow-to-pipe, (85203): 14
90-degree elbow, squaring a, (85203): 10
90-degree elbow-to-pipe alignment, (85203): 10–11
pipe-to-pipe using couplings, (85203): 14–15
basics, (85203): 6
gap setting methods
measure and move, (85203): 5
shimming, (85203): 5
pipefitter responsibilities, (85203): 6
pipe length calculations
center-to-center method, (85203): 5, 10
center-to-face method, (85203): 5–6, 11
face-to-face method, (85203): 6, 11
takeouts, (85203): 8
slip-on sleeve joints, (85203): 16–17
welded valves, (85203): 15–16
Socket weld piping drawings
bill of materials, (85203): 8
center-to-center, (85203): 10
double-line drawings, (85203): 3
elevation view, (85203): 7
plan view, (85203): 7
fitting symbols, (85203): 4–5, 8
isometric drawings, (85203): 4, 8
single-line drawings, (85203): 3
elevation view, (85203): 7
plan view, (85203): 7
Socket welds, (85201): 2, 15, (85203): 7
Sockolet, (85203): 3, 4
Soft (annealed) copper pipe, (85201): 3
Soft metal gaskets, (85207): 46
Soft tubing cutters, (85206): 9
Soldering fluxes, (85204): 9
Solid core PVC (polyvinyl chloride) pipe, (85206): 6
Solid metal gaskets, (85207): 43

Solid wall, (85206): 6, 17
Solvent weld, (85206): 8, 17
Specific notes, (85208): 4
Spiral-wound metal gaskets, (85207): 44
Split-body ball valves, (85207): 7, 8
Spool, (85208): 10, 29
Spool drawings, (85208): 17, 20
Spring-loaded relief valve, (85207): 19
Spur-gear operators, (85207): 24, 25
Stainless steel, oxyacetylene brazing, (85204): 21
Stainless steel pipe, (85201): 2–3, 6
Steam piping systems, (85201): 8–9
Stop valves
    ball valve, (85207): 6–8, 51
    gate valve, (85207): 1–5, 51
    knife gate valve, (85207): 5–6
    plug valve, (85207): 8–10, 51
    symbols, (85207): 34
    three-way valves, (85207): 11, 12
Straight pipe cage clamps, (85202): 18–20
Straight tee, (85203): 3, 21
Strap wrench, (85205): 15
Swage, (85203): 3, 4, 21
Swing check valve, (85207): 20–21, 23
Symbol legends, (85208): 3
Symbols
    butt weld valves, (85202): 9
    fittings, (85202): 8, (85208): 33–34
    pipe hangers and supports, (85208): 22
    specials, (85208): 35
    valves, (85202): 8, (85208): 35

T

Tack-weld, (85203): 7, 21
Takeouts
    butt weld fittings, (85202): 11–13
    caps, (85202): 12
    defined, (85205): 9, 19
    45-degree elbows, (85202): 11–12
    long-radius 90-degree elbows, (85202): 11–13
    180-degree returns, (85202): 12
    short-radius 90-degree elbows, (85202): 11–13
    socket weld fittings, (85203): 5, 8
    threaded pipe, (85205): 9
Taper-taper joints, Bondstrand® GRE pipe, (85206): 1, 2
Tee branch connections, (85202): 3
Tees
    butt weld, (85202): 1–2
    function, (85203): 1
    PVC/CPVC pipe, (85206): 6
    socket weld, (85203): 4
    threaded pipe, (85205): 5
Tee-to-pipe alignment, (85202): 27
Teflon®, (85205): 13–14
Tell-tale holes, (85202): 2
Thermal beveling, (85202): 6, 9
Thermal expansion stress
    calculating, (85201): 9–10
    causes, (85201): 9
    coefficient of expansion by type of, (85201): 10
    compensating for
        cold springing, (85201): 11–12
        expansion joints/loops, (85201): 10–11
        flexibility in layout, (85201): 10, 11
    results, (85201): 9
Thermal transients, (85207): 3, 51
Thermoplastic, (85206): 6, 17
Thread angle, (85205): 1, 19

Thread crest, (85205): 1, 19
Threaded flange, (85207): 37
Threaded pipe fabrication
    hand engagement, (85205): 2, 16
    joint compound, (85205): 13, 14–15
    joint tape, (85205): 13–14
    pipe length calculations
        center-to-center method, (85205): 9–12
        center-to-face method, (85205): 12–13
        face-to-face method, (85205): 13
        makeups, (85205): 9, 11
        takeouts, (85205): 9
    threaded valves, (85205): 16–17
    wrenches, (85205): 15
    wrench makeup, (85205): 2, 16
Threaded pipe fittings
    banded, (85205): 3
    branch connections, (85205): 4–6
    caps, (85205): 6
    cast iron, (85205): 3
    crosses, (85205): 5
    elbows, (85205): 4–5
    flanges, (85205): 8
    forged steel, (85205): 4
    line connections
        couplings, (85205): 6
        reducing bushings, (85205): 7
        reducing coupling, (85205): 6
        unions, (85205): 7, 8
    malleable iron, (85205): 3–4
    nipples, (85205): 7–8, 9
    offsets, (85205): 4–5
    plugs, (85205): 6
    return bends, (85205): 4–5
    tees, (85205): 5
Threaded piping systems, (85205): 1
Threaded valves, (85205): 16–17
Thread engagement, (85205): 9, 19
Thread markings, valves, (85207): 33
3-4-5 method, squaring angles using, (85202): 25
Three-way valves, (85207): 11, 12
Throttling, (85207): 1, 52
Tiles on drawings, (85208): 1
Title block information
    approved by, (85208): 2
    checked by, (85208): 2
    drawing title, (85208): 1
    drawn by, (85208): 2
    key plan, (85208): 2
    project number, (85208): 1
    revision number, (85208): 1–2
    scale, (85208): 1
    sheet number, (85208): 1
    sheet size, (85208): 1, 2
    zone, (85208): 2
Tongue and groove flanges, (85207): 41
Top-entry ball valves, (85207): 7, 8
Torque, (85207): 7, 52
Transverse frame
    for compartment identification, (85208): 7
    defined, (85208): 29
    function, (85208): 6
Transverse frame line, (85208): 6, 29
Trim, (85207): 7, 52
Trim identification markings, (85207): 32
Tubing cutters, (85204): 7, (85206): 9
Turbulence, (85207): 14, 52

Two-flange butterfly valve, (85207): 16, 17
Type ACR copper pipe
    applications, (85201): 3
    characteristics, (85201): 3–4
    color code, (85201): 5
    labeling, (85201): 5
    sizing, (85201): 4
Type A metric copper pipe, (85201): 4
Type B metric copper pipe, (85201): 4
Type C metric copper pipe, (85201): 4
Type C backing rings, (85202): 14–16
Type CC backing rings, (85202): 14–16
Type CCC backing rings, (85202): 14–16
Type DWV copper pipe
    applications, (85201): 3
    color code, (85201): 5
    labeling, (85201): 5
    wall thicknesses, (85201): 3
Type K copper pipe
    applications, (85201): 3
    color code, (85201): 5
    labeling, (85201): 5
    wall thicknesses, (85201): 3
Type L copper pipe
    applications, (85201): 3
    color code, (85201): 5
    labeling, (85201): 5
    wall thicknesses, (85201): 3
Type M copper pipe
    applications, (85201): 3
    color code, (85201): 5
    labeling, (85201): 5
    wall thicknesses, (85201): 3

U

U.N. *See* Unless noted otherwise (U.N.)
Unions
    PVC/CPVC pipe, (85206): 6
    socket weld, (85203): 1–2
    threaded pipe fittings, (85205): 7, 8
Universal torch tip, (85204): 6
Unless noted otherwise (U.N.), (85208): 2
U-shaped fittings, (85202): 1
Utility air piping systems, (85201): 7
Utility water piping systems, (85201): 8

V

Valve actuators
    butterfly valve, (85207): 15
    chain operators, (85207): 25
    components, (85207): 27
    control valve, (85207): 17
    function, (85207): 24
    gear operators
        bevel-gear operators, (85207): 24, 25
        spur-gear operators, (85207): 24, 25
        worm-gear operators, (85207): 24–25, 25
    reach rod, (85207): 27–28, 51
    types of
        air motor-driven actuators, (85207): 27
        electric actuators, (85207): 27
        hydraulic actuators, (85207): 27
        pneumatic actuators, (85207): 27
Valve body, (85207): 3, 52
Valves
    application factors, (85207): 31
    common features, (85207): 1

components
    body, (85207): 1, 51
    bonnet, (85207): 3, 51
    disc, (85207): 1, 51
    reach rod, (85207): 24, 27–28, 51
    seat, (85207): 1, 51
control valve, (85207): 17, 20, 21, 51
on cylinders, (85204): 2
directional flow regulation
    ball check valves, (85207): 22, 23
    butterfly check valve, (85207): 22, 23
    check valve, (85207): 19, 51
    foot valves, (85207): 23–24
    lift check valve, (85207): 21–22, 23
    swing check valve, (85207): 20–21, 23
flow regulating
    angle globe valve, (85207): 11, 13–14, 51
    butterfly valve, (85207): 14–16, 51
    diaphragm valves, (85207): 16–17, 18
    globe valve, (85207): 11, 13–14, 51
    needle valves, (85207): 17, 19
    Y-type, (85207): 14
function, (85207): 1
installation considerations, (85207): 29–30
markings
    bridgewall marking, (85207): 29
    common, (85207): 32
    flow direction arrow, (85207): 15, 29
    installation and, (85207): 29
    rating designation, (85207): 32
    schematic symbols, (85207): 32–33
    size designation, (85207): 33
    standards, (85207): 31
    thread markings, (85207): 33
    trim identification, (85207): 32
materials used in, (85207): 1
multiport valves, (85207): 9, 11
pressure- relief valve, (85207): 19, 21, 22, 51
rigging, (85207): 28–29
safety guidelines, storage and handling, (85207): 28
safety valves, (85207): 19, 22
start and stop
    ball valve, (85207): 6–8, 51
    gate valve, (85207): 1–5, 51
    knife gate valve, (85207): 5–6
    plug valve, (85207): 8–10, 51
    three-way valves, (85207): 11, 12
selection considerations, (85207): 30–31
storing, (85207): 28
symbols, (85202): 8, (85208): 35
threaded valves, (85205): 16–17
welded valves, (85203): 15–17
Valve stem
    defined, (85207): 52
    flow regulation, (85207): 1
    nonrising stem, (85207): 5, 6
    outside screw-and-yoke (OS&Y) stem, (85207): 5, 6
    rising stem, (85207): 5, 6
Valve-to-pipe alignment
    butt weld pipe fabrication, (85202): 27–28
    socket weld valves, (85203): 15–17
Valve trim, (85207): 11, 32, 52
Van Stone flanges, (85202): 2–3
Vendor drawings, (85208): 17, 21
Venturi-type ball valve, (85207): 7
Vinyl gaskets, (85207): 45
Visible (object) lines, (85208): 3

Viton® gaskets, (85207): 45–46
Viton® O-ring, (85203): 1
V-port plug, (85207): 14

W

Wafer lug butterfly valves, (85207): 16, 17
Wafer-type butterfly valve, (85207): 16
Water hammer, (85201): 8, 15
Water line, (85208): 6, 29
Water piping systems, (85201): 8–9
Wedges
   defined, (85207): 52
    flexible, (85207): 2, 3
    solid, (85207): 2, 3
    split, (85207): 2, 3–4
Weep-holes, (85202): 2
Welded valves, (85203): 15–17
Welding goggles, (85204): 19
Welding rings. *See* Backing rings
Weld-neck flanges, (85202): 2, 4, (85207): 35
Weldolets, (85202): 2, 3
Wetting, (85204): 9, 25
Whitworth threads, (85205): 1
Wire drawing, (85207): 13, 52
Witness mark, (85206): 12, 17
Working life, (85206): 2, 17
Worm-gear operators, (85207): 24–25, 25

Y

Yoke bushing, (85207): 30, 52
Y-type, (85207): 14

Z

Zone on title block, (85208): 2